Quantum Mechanics 2:
The Toolkit

N. J. B. Green

Reader in Chemistry, King's College, London

Series sponsor: **ZENECA**

ZENECA is a major international company active in four main areas of business: Pharmaceuticals, Agrochemicals and Seeds, Speciality Chemicals, and Biological Products.

ZENECA's skill and innovative ideas in organic chemistry and bioscience create products and services which improve the world's health, nutrition, environment, and quality of life.

ZENECA is committed to the support of education in chemistry and chemical engineering.

OXFORD   NEW YORK   TOKYO
OXFORD UNIVERSITY PRESS
1998

Oxford University Press, Great Clarendon Street, Oxford OX2 6DP

Oxford New York
Athens Auckland Bangkok Bogota Buenos Aires Calcutta
Cape Town Chennai Dar es Salaam Delhi Florence Hong Kong Istanbul
Karachi Kuala Lumpur Madrid Melbourne Mexico City Mumbai
Nairobi Paris São Paolo Singapore Taipei Tokyo Toronto Warsaw

and associated companies in
Berlin Ibadan

Oxford is a trade mark of Oxford University Press

Published in the United States
by Oxford University Press Inc., New York

A catalogue record for this book is available from the British Library

Library of Congress Cataloging in Publication Data
(Data available)

ISBN 0 19 850227 3 (Pbk)

Typeset by the author

Printed in Great Britain on acid free paper by
The Bath Press, Bath

Series Editor's Foreword

Oxford Chemistry Primers are designed to provide clear and concise introductions to a wide range of topics that may be encountered by chemistry students as they progress from the freshman stage through to graduation. The Physical Chemistry series aims to contain books easily recognized as relating to established fundamental core material that all chemists need to know, as well as books reflecting new directions and research trends in the subject, thereby anticipating (and perhaps encouraging) the evolution of modern undergraduate courses.

In this Physical Chemistry Primer, and its companion volume, Nick Green presents a clearly written and elegant account of the foundations and applications of Quantum Mechanics. He develops and builds on ideas that undergraduate chemists will have encountered via introductory courses to give insight and understanding of the behaviour of the microscopic world at the atomic and molecular level. This Primer will be of interest to all students of chemistry and their mentors.

Richard G. Compton
Physical and Theoretical Chemistry Laboratory
University of Oxford

Preface

Chemistry is the science of molecules and their transformations. Our current understanding of matter at this level is founded on the theory of quantum mechanics. This theory provides our basic picture of atomic and molecular structure, the nature of chemical bonding and the interactions of molecules with one another (intermolecular forces and chemical reactions) and with external fields and radiation (spectroscopy). The influence of quantum mechanics is not simply limited to the parts of chemistry normally identified as 'physical'—it is important both in inorganic chemistry (symmetry, structure, and spectroscopy) and in organic chemistry (orbital symmetry and mechanism). In addition, modern coherent spectroscopic methods, particularly NMR techniques, are essentially quantum mechanical in nature. It is important, therefore, that chemists should have a good grounding in quantum mechanics and this is recognized in most undergraduate courses.

This primer is the companion to the Oxford Chemistry Primer *Quantum mechanics 1: foundations*, which deals with the mathematical formulation of quantum mechanics and the few simple cases where exact solutions can be obtained. It also introduces the methods of group theory, which are used to simplify complex multi-particle problems in the electronic structure of molecules and their energy levels.

Unfortunately most quantum mechanical problems in chemistry can not be solved exactly, so approximate methods must be used. The purpose of this primer is to describe and illustrate the application of these methods, which are used for the formulation and optimization of approximate wavefunctions. Illustrations are given, describing how approximations may be refined, and how states respond to external perturbations, such as magnetic and electric fields. There are so many important applications that a book this length can not cover them exhaustively, so in several places reference is given to longer and more advanced works.

The development in this primer uses many results derived in *Quantum mechanics 1*, cross-references are given wherever they may be helpful.

London
1998

N.J.B.G.

Contents

1 Exact symmetries

In *Quantum mechanics 1* we introduced the idea of molecular symmetry, based on the equilibrium geometry of a molecule and showed how the molecular point group could be applied, both to *classify* the symmetries of molecular vibrations and orbitals, and to *construct* the appropriate wavefunctions using symmetry-adapted linear combinations. However, the molecular point group is not the only type of symmetry group of relevance in chemistry. The molecular point group symmetry is only an approximate symmetry, because the whole idea of a molecular geometry relies on the Born–Oppenheimer approximation (see *Quantum mechanics 1*, Section 2.5). The Schrödinger equation of a molecule is also invariant to a number of other symmetries, which are exact, in the sense of not being dependent on the Born–Oppenheimer approximation:

- translational symmetry
- rotational symmetry
- inversion symmetry
- time-reversal symmetry
- permutation symmetry

1.1   Symmetries of space

Translational and rotational symmetry

These symmetries arise from the uniformity of space. The properties of an object do not depend on where it is situated or how it is oriented in the coordinate system (or equivalently, on the location and orientation of the coordinate system relative to the centre of mass). In classical physics these observations lead to conservation laws for the momentum and the angular momentum. In quantum mechanics the corresponding operators commute with the Hamiltonian, so it is possible to find states with fixed values of the energy, the momentum, and the allowed angular momentum eigenvalues. The eigenfunctions for momentum and angular momentum have already been described in Chapter 3 of *Quantum mechanics 1*.

Inversion symmetry – parity

Inversion symmetry, however, is a different matter. It has no classical counterpart, and it involves the transformation of the coordinates of all the particles (nuclei and electrons) into values inverted through the origin. This inversion operation should not be confused with the inversion operation of the molecular point group. The inversion of the coordinates described here may

The parity of a wavefunction is its symmetry to an inversion of all coordinates through the origin.

alter the configuration of particles so that it can not be superimposed on the original configuration; but it does not alter any of the terms in the Schrödinger equation, and in this sense it is a symmetry. Furthermore, the position of the origin for the inversion is arbitrary: it does not have to be at the centre of mass of the molecule.

This invariance of the Schrödinger equation is easily verified. Inversion leads to the transformation of all position vectors $\mathbf{r}$ to $-\mathbf{r}$. The transformation has no effect on the kinetic energy operator, which only contains second derivatives with respect to the coordinates. Although all first derivatives change sign on inversion, second derivatives do not, since the sign change of the second differentiation cancels out that of the first. Similarly, inverting the coordinates does not change any of the inter-particle distances, so the potential energy operator is also unaffected. Hence the inversion operator has no effect on the Hamiltonian, commutes with it, and does not alter the energy. A consequence of this commutation is that it is possible to contruct a complete set of eigenfunctions of $\hat{H}$ that are also eigenfunctions of the inversion.

Since two successive inversions are equivalent to the identity operation, transforming all wavefunctions into themselves, the only possible eigenvalues for the inversion operator are the square roots of 1, that is ± 1. Wavefunctions can therefore be labelled $+$ or $-$ according to this eigenvalue, which is known as the *parity* of the state.

The parity of a state is very useful in calculating matrix elements or selection rules. For example, two wavefunctions of different parity must be orthogonal, because the integrand of $<+|->$ is odd with respect to inversion. (For an integral to be non-zero, the integrand must be totally symmetric with respect to *all* symmetries, see Section 4.8 of *Quantum mechanics 1*.) Similarly, a matrix element of a scalar quantity must be zero if the bra and the ket have different parities, because a scalar quantity and its corresponding operator (such as energy) are unaltered by inversion; thus matrix elements of the form $<-|+|+>$ or $<+|+|->$ must be identically zero because the parity of the integrand is odd. On the other hand vector quantities (e.g. position, momentum, and dipole moment) have odd parity. Thus all matrix elements of these operators between states of the same parity, that is, of the types $<+|-|+>$ and $<-|-|->$, must be zero. Since the intensity of an optical absorption or emission is proportional to the square modulus of the matrix element of the dipole moment operator, we can conclude that all one-photon $+ \leftrightarrow +$ and $- \leftrightarrow -$ transitions are forbidden.

Time-reversal symmetry

Time-reversal symmetry has deep consequences in quantum mechanics, but we shall defer any discussion of the role of time until Chapter 5.

1.2 Permutation symmetry – the Pauli principle

Permutation symmetry also has no counterpart in classical mechanics, but it has the most important consequences in chemistry, for example the Pauli exclusion principle, which dictates the structure of the periodic table.

Problem 1.1.1. *Show that inversion transfoms all momenta $\mathbf{p}$ to $-\mathbf{p}$, but that angular momenta (defined as $\mathbf{r} \wedge \mathbf{p}$) do not change sign under inversion. Hence, the angular momentum operator is invariant to inversion and it is possible to contruct a complete set of angular momentum eigenfunctions that are also eigenfunctions of the inversion.*

Some nuclear processes appear to violate inversion symmetry, notably β decay. The effect is generally unimportant in chemistry, but inclusion of parity non-conservation in the Schrödinger equation leads to very small energy differences between enantiomers, and could lie at the origin of the biological preference for particular chiralities of amino acid, and DNA.

The Schrödinger equation is invariant to any permutation, or relabelling of identical particles.

A permutation is a relabelling of indistinguishable particles. It is obvious that the observable properties of a helium atom (for example) should not depend on which electron is labelled 1 and which is labelled 2. A quick consideration of the electronic Hamiltonian in atomic units (see p. 27, *Quantum mechanics 1*)

$$-\frac{1}{2}\nabla_1^2\psi - \frac{1}{2}\nabla_2^2\psi - \frac{2}{r_1}\psi - \frac{2}{r_2}\psi + \frac{1}{r_{12}}\psi = E\psi \qquad (1.1)$$

shows that both the kinetic energy operator and the potential energy operator are unchanged by such a relabelling so that it is possible to label each eigenfunction with an eigenvalue of the electron permutation operator. (This conclusion is not altered by inclusion of the nuclear kinetic energy.)

Just as for the inversion operator, when the permutation of two particles is performed twice in succession the result is the identity. Thus the eigenvalues of such a permutation may only be ± 1.

Larger atoms and molecules contain more than two electrons and so there are more complicated reordering permutations to consider. For example in a lithium atom there are $3! = 6$ distinct permutations of the electrons: 123, 132, 213, 321, 231, and 312, as depicted in Fig. 1.1. The first is the identity, the next three are the result of single pair exchanges leaving the third particle unaltered, and the final two can be obtained as the result of two such exchanges. In general, for an atom or molecule with n electrons there are $n!$ such permutations to consider.

All permutations can be expressed as a sequence of simple pair exchanges, and permutations can be divided into two sets, those involving an odd number of exchanges, and those involving an even number. They are called odd and even permutations respectively. It is impossible to construct an even permutation from an odd number of exchanges or vice versa. For three electrons the permutations 123, 231, and 312 are even, and the permutations 132, 213, and 321 are odd. There are always equal numbers of even and odd permutations.

If we consider the $n!$ possible permutations of n identical particles as symmetry operations we find that they form a symmetry group, called a permutation group and given the name S_n. The symmetry group S_3 is isomorphic with the point group C_{3v}, which means that it has the same group multiplication table. The fact that the symmetry operations of C_{3v} induce all possible rearrangements of the three H atoms of an ammonia molecule is a reflection of this fact, which is illustrated by comparing the group multiplication table of C_{3v} in Table 1.1 with that of S_3 in Table 1.2. The permutations are identified by their effect on the ordering 123.

Following the arguments of *Quantum mechanics 1*, Chapter 4, we conclude that the possible wavefunctions of a molecule must generate irreducible representations (irreps) of the electron permutation group. Simultaneously they must also generate irreps of the nuclear permutation group for all collections of identical nuclei in the molecule, for example in the ethene molecule this group will contain the permutation of the two C atoms and all possible permutations of the four H atoms.

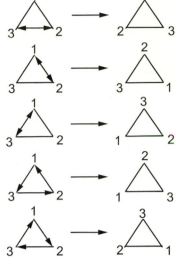

Fig. 1.1 The possible permutations of three identical objects. The identity has been omitted.

Every permutation can be characterized unambiguously as either even or odd.

Table 1.1 C_{3v} group multiplication table – BA

First operation A						
B	E	C_3	C_3^2	σ_1	σ_2	σ_3
E	E	C_3	C_3^2	σ_1	σ_2	σ_3
C_3	C_3	C_3^2	E	σ_2	σ_3	σ_1
C_3^2	C_3^2	E	C_3	σ_3	σ_1	σ_2
σ_1	σ_1	σ_3	σ_2	E	C_3^2	C_3
σ_2	σ_2	σ_1	σ_3	C_3	E	C_3^2
σ_3	σ_3	σ_2	σ_1	C_3^2	C_3	E

Table 1.2 S_3 group multiplication table – BA

	First operation A					
B	123	231	312	132	321	213
123	123	231	312	132	321	213
231	231	312	123	321	213	132
312	312	123	231	213	132	321
132	132	213	321	123	312	231
321	321	132	213	231	123	312
213	213	321	132	312	231	123

All permutation groups have two special irreps – the totally symmetric irrep, in which all the characters are unity, and the *alternating* irrep, in which all even permutations have characters of unity and all odd permutations have characters of minus one. The fact that there are equal numbers of even and odd permutations follows from the orthogonality of these irreducible representations.

The Pauli principle

All particles can be classified as either fermions or bosons. The wavefunction must change sign on the permutation of any pair of identical fermions, and remain invariant to any interchange of identical bosons.

The *Pauli principle* is of central importance in chemistry. According to this principle, elementary particles can be divided into two types, depending on the value of their spin quantum number: *bosons* have integer spins (0, 1, 2, etc.) and *fermions* have odd half-integer spins ($\frac{1}{2}$, $\frac{3}{2}$, etc.) The Pauli principle states that the wavefunction must be antisymmetric to the interchange of any pair of identical fermions and symmetric to all interchanges of identical bosons. In consequence, all odd permutations of fermions must have characters of -1 in the permutation group, and all even permutations of fermions (and all permutations of bosons) must have characters of $+1$. Therefore the only irreps of the permutation groups that can be found in nature are the totally symmetric irrep, for permutations of bosons, and the alternating irrep, for permutations of fermions.

The fact that only these permutation symmetries occur in nature places severe restrictions on the possible forms of the wavefunctions of molecules, with respect to the permutation of electrons and of nuclei.

As a simple illustration we take a two-electron system, such as the helium atom or the hydrogen molecule. If we neglect spin–orbit coupling, the electronic wavefunction can be separated into the product of an orbital function, which depends only on the spatial coordinates of the electrons, and a spin function, which depends only on the spin coordinates (see *Quantum mechanics 1*, p. 31):

$$\psi_{\text{el}} = \phi_{\text{orb}}(\mathbf{r_1}, \mathbf{r_2}) \times \chi_{\text{spin}}(\omega_1, \omega_2) \tag{1.2}$$

We consider the effect of permuting the electrons on each function separately, and then find the symmetry-adapted linear combinations that obey the constraints prescribed by the Pauli principle.

The spin function of a single electron can take only two forms, depending on whether the quantum number m_s is $+\frac{1}{2}$ or $-\frac{1}{2}$. We denote these possibilities respectively $|\alpha >$ and $|\beta >$. In a two-electron system the spin functions can be divided into three sets, which are not interconverted by permutations; each set is therefore a basis for a representation of the electron permutation group S_2. The sets are $\{|\alpha\alpha >\}$ and $\{|\beta\beta >\}$ and $\{|\beta\alpha >, |\alpha\beta >\}$. The first two of these are one-dimensional bases for the totally symmetric irrep, since they are unchanged by permutation of the two electrons. The third set contains two functions that are interconverted by the permutation and so form a two-dimensional basis for the group. The character table of S_2 is given in Table 1.3, together with the representation generated by this basis, labelled Γ. The group has only two irreps: the totally symmetric irrep, A, and the alternating irrep, B. The representation can be reduced by inspection to $A + B$ (alternatively use the reduction formula, eqn 4.28 in *Quantum mechanics 1*). Use of projection operators gives the combinations $|\alpha\beta > +|\beta\alpha >$ for the irrep A and $|\alpha\beta > -|\beta\alpha >$ for the irrep B. The first of these combinations is even with respect to the permutation, and the second is odd.

The overall symmetry of ψ_{el} with respect to permutation of the electrons must be B, and since ψ_{el} is obtained by multiplying the spin function and the orbital function, this symmetry must be the direct product of the spin symmetry with the orbital symmetry (see Section 4.9 of *Quantum mechanics 1*). The direct product table for S_2 is given in Table 1.4, from which we see that either the spin symmetry must be A and the orbital symmetry B, or vice versa.

We next consider the orbital symmetry, and identify two possibilities. Either the two electrons occupy the same orbital, in which case they both have the same orbital function ϕ_1, or they occupy different orbitals. We denote the first of these possibilities $|11 >$, which clearly has A symmetry; the second possibility gives the two functions $|12 >$ and $|21 >$, which are interconverted by the permutation, and so, repeating the analysis just performed for the spin functions, they give the symmetry-adapted linear combinations $|12 > +|21 >$ with A symmetry and $|12 > -|21 >$ with B symmetry.

We therefore conclude that A orbital functions, such as $|11 >$ or $|12 > +|21 >$, which are even with respect to the permutation, can only be combined with the B spin function $|\alpha\beta > -|\beta\alpha >$; this is called a *singlet* state, because of the spin degeneracy of 1 (only the single B spin function is possible). On the other hand, the B orbital function $|12 > -|21 >$ can be associated with any of the three spin functions with A symmetry, $|\alpha\alpha >$, $|\beta\beta >$, or $|\alpha\beta > +|\beta\alpha >$; this state therefore has a spin degeneracy of three, and is called a *triplet* state.

One of the most important results of this type of analysis is that it shows that if the two electrons occupy the same orbital, the spin state must be singlet. It is not possible for two electrons to have simultaneously the same orbital function and the same spin function – this would give an electronic wavefunction with A symmetry, in contravention of the Pauli principle. This argument shows how the familiar *Pauli exclusion principle* is a consequence of the permutation symmetries of the electrons.

Table 1.3 S_2 character table

	12	21
A	1	1
B	1	−1
Γ	2	0

Table 1.4 S_2 direct products

	A	B
A	A	B
B	B	A

The Pauli symmetry can be satisfied for two electrons in two ways: either the orbital function is even and the spin function odd, giving a singlet state, or vice versa, giving a triplet state of degeneracy 3 because of the three possible even spin functions.

Table 1.5 S_3 character table

	123	231, 312	132, 321, 213
A_1	1	1	1
A_2	1	1	−1
E	2	−1	0
(a)	1	1	1
(b)	3	0	1
(c)	6	0	0

Table 1.6 S_3 and C_{3v} direct product table

	A_1	A_2	E
A_1	A_1	A_2	E
A_2	A_2	A_1	E
E	E	E	$A_1 + A_2 + E$

Problem 1.2.1.   *Show that the three-dimensional representation generated by the basis functions $|112>$, $|121>$, and $|211>$ can be reduced to $A_1 + E$.*

Problem 1.2.2.   *Show that this representation can be reduced to $A_1 + A_2 + 2E$.*

The analysis we have just presented could easily have been done without recourse to group theory. However, in more complicated situations group theory enables us to draw conclusions about electronic permutation symmetries and degeneracies much more easily than commonsense. We illustrate this point by considering the possibilities for a three-electron system, such as the Li atom.

The permutation group S_3 is isomorphic with C_{3v}, and its character table is given in Table 1.5. The class of rotational operations in C_{3v} corresponds to the class of *cyclic permutations* in S_3 (imagine the three numbers at the corners of a triangle and rotate the triangle 120°, as shown in Fig. 1.1). Similarly the class of reflections corresponds to the class of simple *pair interchanges* (imagine the effect on the same triangle).

The Pauli principle states that the overall electronic symmetry must be A_2 (the alternating irrep). The direct product Table 1.6 shows that this can only happen if the spin function is A_1 and the orbital function A_2 (or vice versa), or if both have E symmetry, and in the latter case, only one of the four possible product functions will be permissible.

In our three-electron system we can identify three conceivable types of electron configuration:

- all the electrons occupy the same orbital (ϕ_1), $|111>$
- two electrons occupy ϕ_1 and the third is in ϕ_2, e.g. $|112>$
- all three electrons occupy different orbitals, e.g. $|123>$.

In case (a) all permutations give back the same function, which must therefore have A_1 symmetry. Of course this configuration is forbidden by the exclusion principle, but it is instructive to see how this is shown by the group theory analysis. Case (b) gives three possible basis functions, $|112>$, $|121>$, and $|211>$. The cyclic permutations interconvert these basis functions, and therefore have a character of zero. However, for each pair exchange, one basis function is unchanged, giving a character of one.

In case (c) there are six basis functions corresponding to the six different possible orderings of three numbers. Each permutation (apart from the identity) interconverts the basis functions – thus the characters of all permutations (apart from the identity) in this representation are zero. The characters of all three representations are tabulated in Table 1.5, and the results are tabulated in Table 1.7.

Table 1.7   Symmetries of orbital functions in a three-electron atom

Basis functions	Symmetry
$\|111>$	A_1
$\|112>$, $\|121>$, $\|211>$	$A_1 + E$
$\|123>$, $\|231>$, $\|312>$, $\|132>$, $\|321>$, $\|213>$	$A_1 + A_2 + 2E$

Table 1.8 Symmetries of spin functions in a three-electron atom

M_S	Basis functions	Symmetry
$\frac{3}{2}$	$\lvert\alpha\alpha\alpha>$	A_1
$\frac{1}{2}$	$\lvert\alpha\alpha\beta>$, $\lvert\alpha\beta\alpha>$, $\lvert\beta\alpha\beta>$	$A_1 + E$
$-\frac{1}{2}$	$\lvert\alpha\beta\beta>$, $\lvert\beta\alpha\beta>$, $\lvert\beta\beta\alpha>$	$A_1 + E$
$-\frac{3}{2}$	$\lvert\beta\beta\beta>$	A_1

Turning to the spin functions, there are only two possibilities for an electron spin, α and β, giving combinations analogous to cases (a) and (b) above. Combinations analogous to case (c) cannot arise because they would require the existence of three different spin functions. The two functions with values of the M_S quantum number of $\pm\frac{3}{2}$ ($\lvert\alpha\alpha\alpha>$ and $\lvert\beta\beta\beta>$) are separately bases for A_1. The set of three spin functions with two spins up and one spin down ($M_S = \frac{1}{2}$), $\lvert\alpha\alpha\beta>$, $\lvert\alpha\beta\alpha>$, and $\lvert\beta\alpha\alpha>$ are bases for $A_1 + E$, as in case (b) above. (Similarly for the corresponding set with $M_S = -\frac{1}{2}$.) There are thus four spin functions of A_1 symmetry, one with each possible value of the M_S quantum number from $-\frac{3}{2}$ to $+\frac{3}{2}$. These are the four components of a state with spin degeneracy 4 and total spin quantum number $S = \frac{3}{2}$, known as a *quartet* state. Similarly, the E spin functions are only found with M_S values of $+\frac{1}{2}$ and $-\frac{1}{2}$. E functions have a spin degeneracy of 2 and a total spin quantum number of $S = \frac{1}{2}$, and are known as a *doublet* state. These results are summarized in Table 1.8.

Having determined the possible symmetries of the spin functions and the orbital functions, we use the Pauli principle to decide which combinations are possible, that is, which combinations can lead to the only permitted final symmetry A_2.

In electron configuration (a) the orbital function has A_1 symmetry, and can only give the correct overall symmetry in combination with an A_2 spin function. There is no spin function with A_2 symmetry. We therefore conclude that this configuration is impossible; in other words, three electrons may not occupy the same orbital. (Of course this is a statement of the Pauli exclusion principle.)

In configurations of type (b) there is another A_1 orbital function, $\lvert 112 > + \lvert 121 > + \lvert 211 >$, which is forbidden for the same reason (although we could not have worked this out easily from commonsense). There is also a doubly degenerate pair of E orbital functions, which can only be associated with a pair of spin functions of E symmetry, that is, $S = \frac{1}{2}$. This is the case for the ground state Li atom, which has the electron configuration $1s^2 2s^1$. We conclude that only a doublet state is possible for this electron configuration. However, there are further restrictions of the possible combinations of spin and orbital functions in this case. There are two orbital functions of E symmetry, and two spin functions of E symmetry. There are therefore four possible combinations of them, which form a direct product basis for the representation $A_1 + A_2 + E$. The only electronic wavefunction permitted by

Problem 1.2.3. *Use projection operators to find the orbital and spin functions that act as bases for the E irrep. Take the antisymmetrized direct product to show that the only doublet state with $M_S = \frac{1}{2}$ is $\lvert 112 > (\lvert\alpha\beta\alpha > - \lvert\beta\alpha\alpha >) + \lvert 121 > (\lvert\beta\alpha\alpha > - \lvert\alpha\alpha\beta >) + \lvert 211 > (\lvert\alpha\alpha\beta > - \lvert\alpha\beta\alpha >).*

the Pauli principle is the combination with A_2 symmetry, which is the antisymmetrized direct product (see Section 4.9 of *Quantum mechanics 1*).

We would have come to the same conclusion, that the only possible spin function for the configuration is a doublet, if we had ignored the full $1s$ orbital. It is generally true, and simplifies the analysis of larger systems significantly, that for a full subshell, the orbital function is totally symmetric and the spin is singlet. We therefore obtain the correct spin degeneracies and orbital symmetries if we only consider electrons in partly occupied orbitals. We can see how this is true for the Li atom by noting that the electronic wavefunction with $M_S = \frac{1}{2}$, is a superposition of three configurations, each of which corresponds to two electrons in orbital 1 with a singlet wavefunction, and the third electron in orbital 2 with spin up.

Problem 1.2.4. *Find the orbital wavefunction for the three 2p electrons in the quartet state of the N atom, in terms of the basis functions $|123>$ etc.*

In case (c) the A_2 orbital function must combine with the A_1 (quartet) spin functions, and the two pairs of E orbital functions, go with the doublet, E, spin functions. We therefore obtain a quartet and two doublet states. The ground configuration of the N atom is an example of such a case, in which (ignoring full subshells) there are three $2p$ electrons occupying different orbitals.

1.3   Determinants

The orbital approximation can be improved to fulfil the Pauli symmetry by constructing the wavefunction as a Slater determinant.

According to the Pauli principle every electronic wavefunction of an atom or a molecule must be a basis for the alternating irrep of the electron permutation group. The wavefunction is constructed from *spin–orbitals*, which are products of one-electron orbital wavefunctions and spin functions. For example, the spin–orbital $|1s> |\alpha>$ specifies that the electron is in a $1s$ orbital with spin up. For any given configuration we can write down a wavefunction in the orbital approximation by associating each electron with a spin–orbital and multiplying them together. Every permutation of electrons will generate another such function, and the total electronic wavefunction is an antisymmetrized combination of these basis product functions. This can be done using the projection operator for the alternating irrep, which will multiply every permutation by ± 1 depending on whether the permutation is even or odd.

However, there is an equivalent way of constructing such an antisymmetrized product, known as a determinant. The idea can be stated as follows. In a configuration there is the same number of spin–orbitals as electrons. Spin–orbitals are arranged in a determinant so that the first column contains a list of the possible spin–orbitals expressed as a function of the coordinates of electron 1, the second column contains the same list for electron 2, and so on. Explicitly, the functions in column one are functions of the coordinates of electron 1, and so on. The first row of the determinant thus contains spin–orbital function one, expressed as a function of the coordinates of each electron in turn. It can easily be seen that such a determinant has the required permutation symmetry. Interchange of two electrons amounts to interchange of two columns of the determinant, and this operation changes the sign of the determinant. An even number of swaps brings the determinant back to its original value, and an odd number changes its sign.

As an illustration we reconsider the case of ground state Li. One of the allowed configurations puts electron 1 in the $1s$ orbital with spin up, electron 2 in $1s$ with spin down, and electron 3 in $2s$ with spin up. It is conventional to distinguish a spin–orbital with spin down by placing a bar over the orbital label. The Slater determinant then becomes

$$\begin{vmatrix} 1s(1) & 1s(2) & 1s(3) \\ \overline{1s}(1) & \overline{1s}(2) & \overline{1s}(3) \\ 2s(1) & 2s(2) & 2s(3) \end{vmatrix} \tag{1.3}$$

which can be expanded to give the combination

$$|1\bar{1}2> - |12\bar{1}> + |\bar{1}21> - |\bar{1}12> + |21\bar{1}> - |2\bar{1}1>$$

This is identical to the result of problem 1.2.3.

Any attempt to put two electrons in the same spin–orbital leads to two identical rows in the determinant. Interchange of these rows leads to a change of sign, but the interchanged determinant is indistinguishable from the original determinant, which must therefore be zero.

The situation is not always this simple. If the three electrons had been in different orbitals we could have found valid determinants based on permutations of any of the three product functions $|12\bar{3}>$, $|1\bar{2}3>$, or $|\bar{1}23>$. Furthermore, any linear combination of these determinants is also fully antisymmetrized. One of these linear combinations corresponds to the quartet state described above, and the other two correspond to the two doublet states. To find which is which we need to use the full group-theoretical methods described above.

Exchange interaction

The introduction of Slater determinants complicates matters somewhat in comparison with the simple orbital approximation. However, virtually all modern quantum chemical calculations use product functions antisymmetrized in this way. In addition, the effect of antisymmetrization is to introduce a *correlation* into the spatial distributions of the electrons.

Consider the case of a two-electron system where the electrons occupy distinct orbitals. In the simple orbital approximation the electronic wavefunction is $\phi_1(\mathbf{r}_1)\phi_2(\mathbf{r}_2)$, and the resulting joint probability density function for the positions of the two electrons is

$$\phi_1^*(\mathbf{r}_1)\phi_1(\mathbf{r}_1)\phi_2^*(\mathbf{r}_2)\phi_2(\mathbf{r}_2)$$

The joint probability density function is a product of a function of $\mathbf{r}_1$ and a function of $\mathbf{r}_2$ and so the positions of the two particles are said to be independent.

However, we have already established that only two states are consistent with the Pauli principle – either the orbital function is symmetric and the spin function antisymmetric with respect to permuting the electrons (the singlet state) or else the orbital function is antisymmetric and the spin function symmetric (the triplet state).

In the triplet state the orbital function is

$$[\phi_1(\mathbf{r}_1)\phi_2(\mathbf{r}_2) - \phi_2(\mathbf{r}_1)\phi_1(\mathbf{r}_2)]/\sqrt{2} \tag{1.4}$$

The resulting probability density function is therefore

$$p(\mathbf{r}_1, \mathbf{r}_2) = \frac{1}{2}[\phi_1^*(\mathbf{r}_1)\phi_1(\mathbf{r}_1)\phi_2^*(\mathbf{r}_2)\phi_2(\mathbf{r}_2)$$
$$+ \phi_2^*(\mathbf{r}_1)\phi_2(\mathbf{r}_1)\phi_1^*(\mathbf{r}_2)\phi_1(\mathbf{r}_2)$$
$$- \phi_1^*(\mathbf{r}_1)\phi_2(\mathbf{r}_1)\phi_2^*(\mathbf{r}_2)\phi_1(\mathbf{r}_2)$$
$$- \phi_2^*(\mathbf{r}_1)\phi_1(\mathbf{r}_1)\phi_1^*(\mathbf{r}_2)\phi_2(\mathbf{r}_2)] \qquad (1.5)$$

The first term in the square brackets is the probability density we obtained from the simple orbital approximation, and the second term is the same with the two electrons permuted. The density contains the average of these two distributions. However, there are also cross-terms, correlating the distributions of the two electrons. These terms are entirely non-classical. The wavefunction is a superposition of the two simple product functions, and the cross-terms arise from interference between them. Their effect on the density is dramatic. If we consider the behaviour of the function as the two electrons approach one another, we find that the cross-terms cancel out the classical distribution, in other words the interference between the product functions in the triplet state is destructive. It is therefore not possible to find two electrons at the same position in a triplet state. The electrons tend to keep apart from one another. This is not an effect of the electrostatic repulsion between them, it is simply an effect of the permutation symmetry, and is depicted in Figs 1.2 to 1.4 for two

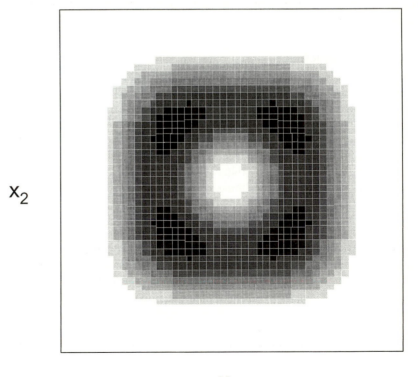

x_2

x_1

Fig. 1.2   Contour plot of the joint density of two non-interacting indistinguishable particles in a one-dimensional box, one in $n = 1$ and one in $n = 2$. No symmetry constraint has been applied. The darker the area the higher the probability density.

non-interacting particles in a box, one in $n = 1$ and the other in $n = 2$. Fig. 1.2 represents the joint probability density of the two x-coordinates if the particles are distinguishable, and there is no symmetry constraint. Fig. 1.3 represents the case where the particles are indistinguishable, spin $\frac{1}{2}$, and in a triplet state. Note the node appearing in the joint probability density along the line $x_1 = x_2$, known as a *Fermi hole*.

Considering the singlet state, we find a very similar expression for the probability density, except that the cross-terms are added rather than subtracted. The interference between the product functions is constructive in the singlet state, and the probability of finding the two electrons in coincidence is therefore twice as large as we would have expected from the simple orbital approximation. This situation is depicted in Fig. 1.4. Note how the density is concentrated around the line $x_1 = x_2$, and the line $x_1 + x_2 = a$ is now a node.

Returning to a two-electron system, we now evaluate the expectation electron–electron repulsion energy for the triplet and singlet functions:

$$< \frac{1}{r} > = \frac{1}{2}\{< 12|\frac{1}{r}|12 > + < 21|\frac{1}{r}|21 >$$
$$\pm [< 12|\frac{1}{r}|21 > + < 21|\frac{1}{r}|12 >]\} \qquad (1.6)$$

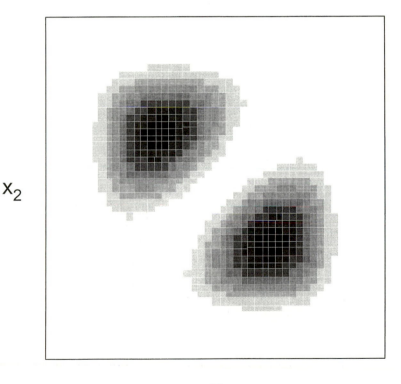

Fig. 1.3 Contour plot of the joint density of two non-interacting indistinguishable particles in a one-dimensional box, one in $n = 1$ and one in $n = 2$. The spatial wavefunction is antisymmetrized with respect to exchange, as in a triplet state. The darker the area the higher the probability density.

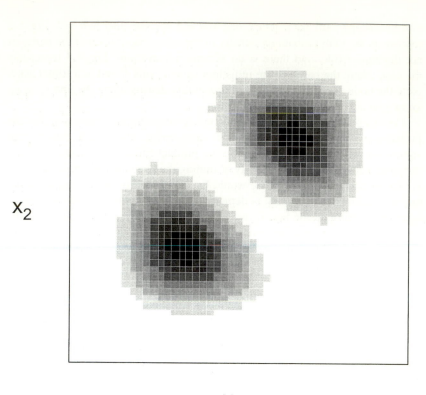

X₂

X₁

Fig. 1.4   Contour plot of the joint density of two non-interacting indistinguishable particles in a one-dimensional box, one in $n = 1$ and one in $n = 2$. The spatial wavefunction is symmetrized with respect to exchange, as in a singlet state. The darker the area the higher the probability density.

Since $\frac{1}{r}$ is unaltered by permuting the two electrons the first and second terms are equal and the third and fourth terms are equal, giving

$$< \frac{1}{r} > \; = \; < 12|\frac{1}{r}|12 > \; \pm \; < 12|\frac{1}{r}|21 > \tag{1.7}$$

The first term is known as the *Coulomb energy*, and is the expectation repulsion energy that would be obtained from the simple orbital approximation. The second term is the *exchange energy* and is the modification to the repulsion energy that arises from the permutation symmetry. The effect of the Fermi hole in the triplet state is to reduce the electron–electron repulsion energy, and thus the triplet state generally lies lower in energy than the corresponding singlet state. A more accurate way of thinking about this is that the tendency of electrons to stay apart in the triplet state permits the orbital functions to shrink closer to the nucleus, thus reducing the energy of the triplet state.

It is possible (but beyond the scope of this book) to derive equations for self-consistent field theory using properly antisymmetrized product functions. The resulting equations are similar to the Hartree equations described in *Quantum mechanics 1*, Section 2.7. The equation for a single spin–orbital turns out to be an eigenvalue equation for an operator that contains the usual one-electron Hamiltonian (kinetic energy and electron-nuclear attraction) and an electron–electron repulsion operator, in which the potential energy of repulsion is averaged over all the other occupied spin–orbitals, just as in the Hartree method. However, in addition, because of the antisymmetrization of the wavefunction there is a third operator, the exchange operator, which represents the contribution of the exchange interaction to the energy of the spin–orbital. The resulting set of eigenvalue equations are known as the Hartree–Fock equations, and are solved iteratively by the self-consistent field method.

The effective Hamiltonian in the Hartree–Fock equations, the Fock operator, is a function of the coordinates of one electron only. For convenience we choose this to be electron 1, and omit all reference to it from the resulting equations:

$$[\hat{H} + \sum_j \hat{J}_j - \sum_j \hat{K}_j]\chi_i = \epsilon_i \chi_i \qquad (1.8)$$

in which χ_i is spin orbital i, $\hat{H}$ is the one-electron Hamiltonian containing all interactions except the electron–electron repulsions, $\hat{J}_j$ is the Coulomb operator,

$$\hat{J}_j = <j|1/r_{12}|j> \qquad (1.9)$$

the integral extending over the coordinates of electron 2. $\hat{K}_j$ is defined

$$\hat{K}_j|i> = <j|1/r_{12}|i>|j> \qquad (1.10)$$

so that it leads to the permutation of electrons 1 and 2; again the integral extends over the coordinates of electron 2.

The Hartree–Fock method is the basis of virtually all modern quantum chemical calculations for atoms and molecules. For a fuller discussion the reader is referred to Szabo and Ostlund (1982).

1.4 Electronic symmetries

We have discussed at length the symmetry of the electronic wavefunction with respect to permutations of the electrons. However, since each electronic wavefunction is an eigenfunction of the electronic Hamiltonian in the Born–Oppenheimer approximation, it is also a basis for an irrep of the molecular point group. The spin is unaltered by any symmetry operation of the point group, so we only need to consider the orbital part of the wavefunction.

The electronic wavefunction is constructed from products of spin–orbitals, each of which carries a symmetry (see *Quantum mechanics 1*, Section 4.3). The basis of all possible products of these functions is therefore a basis for the appropriate direct product representation of the point group. However, the permissible combinations of the orbital functions are prescribed by the Pauli

The Pauli principle prescribes which combinations or orbital and spin functions are permitted. The permutation symmetry must now be combined with the point group to characterize the electronic wavefunction.

principle, as the preceding discussion of electron permutations should make clear.

In spectroscopy it is usual to give an electronic energy level (or *term*) one label denoting its spin degeneracy and another denoting its orbital symmetry, which identifies the irrep of the point group generated by the electronic wavefunction. This label also tells us the orbital degeneracy of the term. Thus, for example, a ^2E term (read doublet E) would have an orbital degeneracy of 2 (E is a two-dimensional irrep) and a spin degeneracy of 2 (doublet spin function) giving a total electronic degeneracy of $2 \times 2 = 4$. We therefore need a method to work out these orbital symmetries.

Closed shell systems

First we should consider the origin of orbital degeneracies. It is not possible to have an orbital degeneracy unless there is more than one symmetrically equivalent choice for at least one orbital in the configuration. Thus, for example, in the B atom the $2p$ electron can occupy any of the three symmetry-related $2p$ orbitals, giving an orbital degeneracy of 3. However, in a closed shell molecule there is no such ambiguity – the set of orbital functions is defined completely by specifying the configuration (in the ground state Be atom, two electrons occupy the $1s$ orbital, and the other two occupy the $2s$ orbital). There is therefore no possibility of orbital degeneracy in a closed shell atom or molecule: the electronic wavefunction must generate a one-dimensional irrep of the point group. Furthermore, since each orbital is doubly occupied, the corresponding orbital function occurs twice in every term of the product. It is not possible for the product to change sign under any symmetry operation, so it must generate the totally symmetric irrep, and hence, for example the atomic term symbol ^1S. This result is very useful, since, as discussed above for the permutation group, in an open shell molecule we only need to consider the electrons in the partly filled orbitals to work out the possible electronic symmetries.

One unpaired electron

In a molecule with one unpaired electron the filled core orbitals make no contribution to the orbital symmetry or the spin, so the orbital symmetry is defined by the symmetry of the singly occupied orbital. The spin state is a doublet. For example, the ground state Li atom $(1s^2 2s^1)$ has the term symbol ^2S, indicating that the electron occupies an orbital with S symmetry. Similarly, the ground state B atom $(1s^2 2s^2 2p^1)$ term symbol is ^2P.

Two unpaired electrons

If a molecule contains two unpaired electrons, the resulting symmetries depend on whether they occupy the same set of degenerate orbitals. If they do not then the possible orbital symmetries are given by the direct product of the orbital symmetries without restriction. The spins can combine either to give symmetric (A) triplet spin functions, which are associated with antisymmetric

(B) orbital combinations such as $|12> - |21>$, or an antisymmetric (B) singlet spin function, which goes with A orbital functions such as $|12> + |21>$. In the spherical point group R_3 the direct product $P \times P = S + [P] + D$, hence the excited state C atom with configuration $1s^2 2s^2 2p^1 3p^1$ has energy levels with the following term symbols: ^1S, ^3S, ^1P, ^3P, ^1D, ^3D; each term will have a different energy as a result of the different exchange interaction in each combination.

The general rule for combining two spins S_1 and S_2 from terms arising from distinct orbitals, such as the excited C atom, is that the resultant spin may take all values from $S_1 + S_2$ to $|S_1 - S_2|$. Thus, for example the product of a doublet and a triplet will give spins of $1 + \frac{1}{2}$ and $1 - \frac{1}{2}$, that is, a quartet and a doublet. The combination of angular momenta will be discussed in more detail at the end of this Section.

If the two electrons occupy the same subshell, as in the ground state C atom ($1s^2 2s^2 2p^2$), the situation is different, because the possibilities are restricted by the Pauli principle. We still have orbital combinations such as $|12> - |21>$ associated with triplet spin functions, but these combinations generate only the antisymmetrized direct product representation. We also have to consider functions such as $|11>$, which are now part of the same configuration, and which, with combinations of the type $|12> + |21>$, generate the symmetrized direct product representation. These must be associated with the singlet spin function. In the case of the ground state C atom, these restrictions give only the states ^1S, ^3P, ^1D.

In practice, working out these term symbols is straightforward. We simply look up the direct product of the symmetries of the orbitals involved; in the first case we apply singlet and triplet spins to each possibility; in the second case we apply singlets to the symmetrized combinations and triplets to the antisymmetrized combinations, which are generally identified in the tables.

For example, we consider the ground state O_2 molecule, which has the configuration $\ldots \pi_g^2$. The direct product table for the $D_{\infty h}$ group gives $\Pi_g \times \Pi_g = \Sigma_g^+ + [\Sigma_g^-] + \Delta_g$. The O_2 ground configuration therefore splits into energy levels with the term symbols $^1\Sigma_g^+$, $^3\Sigma_g^-$, and $^1\Delta_g$. The triplet state is the ground state because of its larger exchange stabilization.

Problem 1.4.1. *Find the allowed term symbols for the following configurations in the point group T_d: e^2, $e^1 t_2^1$, t_2^2.*

Three electrons

The case of three electrons in a doubly degenerate orbital is simple. Only spin functions containing two up and one down (or vice versa) are possible, as discussed already, so we have a doublet spin state. Similarly we must have two electrons in one orbital and one in the other, so that there is an orbital degeneracy of 2 associated with this choice. The orbital symmetry of the state is the same as the symmetry of the degenerate orbital. So, for example, the term symbol for the O_2^- ion (superoxide), whose electron configuration is $\ldots \pi_g^3$, is $^2\Pi_g$.

The case of three electrons in a higher degeneracy is considerably more complicated. We consider in detail only the case of three electrons in a triply degenerate subshell. We can get some help by reference to the permutation

The case of three electrons with a high orbital degeneracy can be conveniently tackled by considering the effects of distortions that remove the degeneracy.

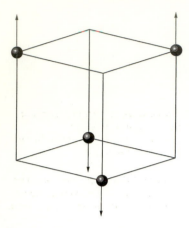

Fig. 1.5   A tetrahedral molecule, showing the direction of a tetragonal distortion.

Table 1.9 T_d descent of symmetry table

T_d	D_{2d}	C_{3v}
A_1	A_1	A_1
A_2	B_1	A_2
E	$A_1 + B_1$	E
T_1	$A_2 + E$	$A_2 + E$
T_2	$B_2 + E$	$A_1 + E$

symmetry analysis performed in the last section, which indicates that configurations where the electrons occupy different orbitals, such as $|123>$, have one combination associated with a quartet spin function and two more associated with doublets. There is no orbital degeneracy to consider in this case since each of the three orbitals must appear in the product function. The other possible configurations are those where two electrons occupy the same orbital, such as $|112>$. For each choice of a singly occupied orbital and a doubly occupied orbital only one combination is permitted, which must be associated with the doublet. There are six possible choices for the combination of two orbitals in such a configuration.

We therefore expect a quartet state, which will be orbitally non-degenerate, and a total of eight orbital functions associated with the doublet spin functions. There is unfortunately no simple method for working out the symmetries of these permitted combinations in the molecular point group.

The most useful method is the *distortion method*, where the effect of a small distortion on the molecule is considered. The distortion is such as to change the point group to one its subgroups, and to lift the degeneracy of the orbitals, wholly or partly. The terms are then worked out for the distorted molecule, then we consider which terms in the original point group could give rise to the terms found for the distorted molecule.

We illustrate the method for a (t_2^3) configuration in a tetrahedral molecule. Initially we consider a tetragonal distortion (if the tetrahedron is constructed on the corners of a cube, the distortion is obtained by stretching the cube along the z axis, see Fig. 1.5). The distorted molecule belongs to the D_{2d} point group. The correlations between irreps of T_d and those of D_{2d} are shown in Table 1.9. It can be seen that the t_2 orbitals split into a b_2 orbital and doubly degenerate e orbitals. The possible electron configurations of the distorted molecule corresponding to t_2^3 are therefore $b_2^2 e^1$, $b_2^1 e^2$, and e^3. The first of these is easy: since b_2^2 is a filled orbital, it has no effect on the symmetry, which is therefore determined by the remaining electron to be ^2E. The final configuration also gives only a ^2E term. The second configuration must be found in two parts. The e^2 configuration would give $^1A_1 + {}^3A_2 + {}^1B_1 + {}^1B_2$ according to the direct product table for D_{2d}, Table 1.10. The direct product of these symmetries with $^2B_2$ gives $^2B_2 + {}^4B_1 + {}^2B_1 + {}^2A_2 + {}^2A_2$, combining the spins according to the vector addition rule discussed above. Checking the correlation Table 1.9, we now recognize that the $^4B_1$ term can only correlate

Table 1.10 D_{2d} direct product table

	A_1	A_2	B_1	B_2	E
A_1	A_1	A_2	B_1	B_2	E
A_2	A_2	A_1	B_2	B_1	E
B_1	B_1	B_2	A_1	A_2	E
B_2	B_2	B_1	A_2	A_1	E
E	E	E	E	E	$A_1 + [A_2] + B_1 + B_2$

with a $^4A_2$ term in the undistorted molecule. The $^2B_2$ term can only be found in combination with a ^2E, correlating with a $^2T_2$ term in the undistorted molecule. Similarly the $^2A_2$ term and the remaining ^2E term must come from a $^2T_1$. This leaves us with two remaining terms, $^2A_1 + {}^2B_1$ to account for. Unfortunately there is an ambiguity at this point – they could correlate with $^2A_1 + {}^2A_2$ or with ^2E in the undistorted molecule.

To resolve this ambiguity we have to consider a second distortion which does not lift the degeneracy of the E irrep. A suitable distortion is to stretch the molecule along one of its threefold axes, reducing the symmetry to C_{3v}; see Fig. 1.6. The correlations can also be found in Table 1.9.

We can recognize some of these correlations from the results we have already obtained from the tetragonal distortion: $^4A_2 \rightarrow {}^4A_2$, $^2T_1 \rightarrow {}^2A_2 + {}^2E$, and $^2T_2 \rightarrow {}^2A_1 + {}^2E$. This leaves us with only a ^2E to account for, which must correlate with ^2E in the undistorted molecule. The required terms of the undistorted molecule are therefore $^4A_2$, ^2E, $^2T_1$, and $^2T_2$. This is virtually the only important case of three electrons in a high degeneracy. The only other important instance is the ground state N atom, which has three $2p$ electrons. As discussed above, in both cases there are eight orbital functions associated with doublet terms.

Problem 1.4.2. *Perform a distortion analysis to give the terms* $^2E + {}^2A_1 + {}^4A_2 + {}^2A_2 + {}^2E + {}^2E$ *for the distorted molecule.*

Problem 1.4.3. *Perform a distortion analysis for the ground configuration of the N atom to find the terms* $^4S + {}^2P + {}^2D$.

Electronic symmetry and angular momentum

The electronic Schrödinger equation for an atom does not include any term containing the spin. In the central field approximation each electron moves in the spherically symmetrical averaged field of the nucleus and the other electrons (*Quantum mechanics 1*, Section 6.1). In such a system the total angular momentum L of all the electronic orbital motions is conserved. Similarly, if the spin and orbital motions are separable (*Quantum mechanics 1*, Section 2.6), the total spin angular momentum S is also conserved.

Although the spin motion does not contribute to the electronic Hamiltonian, the necessity for the electronic wavefunction to be antisymmetric with respect to all permutations of electron pairs limits the possible combinations of spin functions and orbital functions (Section 1.2). The different exchange interactions associated with the different ways in which the Pauli antisymmetrization principle is satisfied lead to a splitting of the energy of each possible electron configuration into terms, each term having particular values of S and L. The number of states in an atomic term arises from the different directions in which each angular momentum vector can point in space relative to the z axis, and is therefore $(2L + 1)(2S + 1)$.

Relativistic corrections to the Schrödinger equation reveal the presence of an interaction between the spin and the orbital angular momenta in an atom, splitting each term into a few closely spaced energy levels. We shall discuss this fine structure in Chapter 4. For the moment we simply consider which terms can arise from a particular electron configuration. First, notice that filled subshells make no contribution to either the orbital or the spin angular momentum, because in every possible combination of spin–orbitals each electron is paired with another, with the same orbital function and the opposite

The term symbol for an atom permits immediate identification of the total orbital angular momentum quantum number L. The analysis of angular momenta gives a more elementary but tedious way to work out the permitted term symbols for a given configuration.

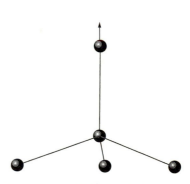

Fig. 1.6 A tetrahedral molecule, showing the direction of a trigonal distortion.

Table 1.11 Angular momenta for p^6

	m_l		M_L	M_S
-1	0	1		
↑↓	↑↓	↑↓	0	0

Table 1.12 Term symbols for p^2

	m_l		M_L	M_S	
-1	0	1			
↑↓			-2	0	^1D
	↑↓		0	0	^1S
		↑↓	2	0	^1D
↑	↑		-1	1	^3P
↑	↓		-1	0	^1D
↓	↑		-1	0	^3P
↓	↓		-1	-1	^3P
↑		↑	0	1	^3P
↑		↓	0	0	^1D
↓		↑	0	0	^3P
↓		↓	0	-1	^3P
	↑	↑	1	1	^3P
	↑	↓	1	0	^1D
	↓	↑	1	0	^3P
	↓	↓	1	-1	^3P

spin, and the angular momentum of a particular orbital is always cancelled by an equal and opposite contribution from another orbital. For example, as shown in Table 1.11, a full p subshell contains two electrons with each possible value of the magnetic quantum number, $m_l = -1, 0, 1$, which represents the component of the orbital angular momentum in the z direction. For every electron with $m_l = +1$ there is another with an equal and opposite $m_l = -1$, yielding a total $M_L = 0$. Similarly there are three electrons with quantum number $m_s = \frac{1}{2}$ and the resultant angular momentum is cancelled out by the three electrons with $m_s = -\frac{1}{2}$, yielding a total $M_S = 0$.

The terms which arise for an atomic configuration can be found by application of group theoretical methods, as discussed earlier in this section. However, because the quantum number L not only acts as a symmetry label on the orbital wavefunction, but also represents the orbital angular momentum of the term, the combinations of L and S permitted by the Pauli principle can also be derived from an analysis of the angular momenta of the possible configurations. The method is to consider all possible arrangements of the electrons consistent with the given configuration and the Pauli exclusion principle. The total $\mathbf{L}$ and $\mathbf{S}$ are vector sums of the individual $\mathbf{l}$ and $\mathbf{s}$ vectors, summed over all the electrons in the atom. Since the individual m_l and m_s represent the components of $\mathbf{l}$ and $\mathbf{s}$ respectively in the z direction, these are added to give the allowable z components of the resultant total orbital and spin angular momentum vectors, $\mathbf{L}$ and $\mathbf{S}$, which are given by the quantum numbers M_L and M_S. Once a set of permitted values of M_L and M_S has been constructed, it simply remains to identify the values of L and S that can give rise to them.

As always, it is much easier to perform the operation than it is to describe it. We shall illustrate the method by considering the ground configuration of the C atom, $1s^2 2s^2 2p^2$, already considered by group theoretical methods on p. 15. Recognizing that the two filled s subshells do not contribute to the angular momentum of the atom, we only need to consider the incomplete $2p$ subshell. The various possible arrangements of the two p electrons consistent with the Pauli principle are tabulated in Table 1.12, together with the corresponding values of M_L and M_S. The largest value of M_L present is 2, so there must be a state present with $L = 2$. The value $M_L = 2$ appears *only* with a spin $M_S = 0$, so the term must have $S = 0$, otherwise other values of M_S would be found in combination with $M_L = 2$. This term therefore has $L = 2$ and $S = 0$, labelled ^1D, and is five-fold degenerate. We have therefore labelled five states with the correct values of M_L and M_S to make up the ^1D term in the table.

The largest value of M_L remaining in the table is $M_L = 1$, and it can be seen that this value appears three times, accompanied by $M_S = 1$, 0, and -1. We can therefore conclude that there is a term with $L = 1$ and $S = 1$, a ^3P term, which contains a total of nine states. Once we have identified nine suitable states in the table there is only one state remaining, which has $M_L = 0$ and $M_S = 0$. It is therefore a ^1S term.

We conclude that the ground electronic configuration of the C atom gives rise to three terms with symbols ^1D, ^3P, and ^1S, in agreement with our earlier analysis. The different energies of these terms arise from the different ways of satisfying the requirement that the total electronic wavefunction be antisymmetric with respect to permuting the two p electrons. The ground term is the ^3P term. This term has a symmetric spin function and an antisymmetric orbital function. Because the orbital function changes sign when $\mathbf{r}_1$ and $\mathbf{r}_2$ are interchanged, it must approach zero when the two particles approach one another, the *Fermi hole*, illustrated in Fig. 1.3. It should be emphasized that this hole is a result of the symmetry of the orbital function, and is additional to any effect of electron–electron repulsion.

It is a general empirical rule, *Hund's rule*, that in the ground configuration the term with the highest value of S lies lowest in energy. This is essentially because the spin function has the maximum symmetry and so the orbital function is as antisymmetric as possible, thus reducing the electron–electron repulsions. In fact the situation is more complicated than this. The reduced electron–electron repulsion permits the orbitals to contract closer to the nucleus, thus reducing the potential energy further, but leading to a subsequent increase in the electron–electron repulsion. However, Hund's rule is generally found to be true.

We complete this subsection by discussing two more general rules which are of great use in constructing the permissible term symbols for a given configuration. First angular momenta from different subshells can be added according to the normal rules for vector addition (the Clebsch–Gordan series, see Section 4.3). The only complications arise as a result of the Pauli principle when we consider electrons in the same subshell. Thus, for example, the excited configuration of the C atom $1s^2\,2s^2\,2p^1\,3p^1$ has all possible combinations of two orbital angular momenta of one unit (i.e. 0, 1 and 2) and all possible combinations of two spins one half (i.e. 0 and 1), giving the terms ^1S, ^3S, ^1P, ^3P, ^1D, and ^3D.

The second useful observation is that there is a symmetry between the terms found for a subshell less than half full and a subshell more than half full. The same terms arise for a subshell containing a given number of electrons and a subshell with the same number of vacancies, or holes. Thus, for example, the O atom, whose ground configuration contains four $2p$ electrons, or two holes in the $2p$ subshell, gives the same terms as the C atom, namely ^3P, ^1D, and ^1S (see Problem 1.4.4).

Problem 1.4.4. *Use the angular momentum method to find term symbols for the ground configurations of the O and N atoms.*

1.5 Nuclear permutation symmetry

In addition to the electron permutation symmetry, which has such important chemical consequences, molecules with identical nuclei must also have symmetry with respect to nuclear permutations. The collection of all permutations of identical nuclei in a molecule forms a group known as the complete nuclear permutation (CNP) group.

To a very good approximation the nuclear spin wavefunction can be separated from the rest of the molecular wavefunction (the rovibronic wavefunction). In consequence, we can consider the symmetries of the nuclear spin functions separately from the rovibronic wavefunction. A nuclear spin degeneracy can then be associated with each molecular energy level. These nuclear spin degeneracies give rise to observable intensity alternation effects in molecular spectra and, in the most extreme cases, to the complete absence of certain energy levels from a molecule, and hence missing lines from the spectrum.

Homonuclear diatomic molecules

The simplest case to consider is that of a homonuclear diatomic molecule, such as H_2. We have already done most of the work in discussing electron permutation symmetries, but there is one important difference: nuclei may be either fermions or bosons, whereas all electrons are fermions. If the two nuclei in a diatomic molecule are fermions (as in H_2) then the total molecular wavefunction must have B symmetry in the CNP group S_2, as in the case of two electrons. If, on the other hand, the two nuclei are bosons (as in N_2), then the total wavefunction must have A symmetry. We consider the wavefunction in two parts: first the contribution of the nuclear spin wavefunction and then the contribution of the rovibronic wavefunction. The symmetry of the overall wavefunction is the direct product of the symmetries of each part.

A nuclear spin I has a space degeneracy of $2I + 1$, corresponding to the $2I + 1$ different possible values for its z component, M_I. Let us denote this degeneracy g_I. In a diatomic molecule there are $g_I(g_I - 1)$ different combinations in which the two M_I values are different (g_I choices for nucleus 1 and then $g_I - 1$ for nucleus 2). There are therefore $\frac{1}{2}g_I(g_I - 1) = (2I + 1)I$ independent antisymmetric nuclear spin functions of the form $\alpha\beta - \beta\alpha$, which have B symmetry. There is the same number of symmetric combinations of the form $\alpha\beta + \beta\alpha$, which have A symmetry. In addition there are g_I combinations in which the two nuclear spins are the same, such as $\alpha\alpha$, and which are automatically symmetric with respect to permutation of the nuclei. In total there are therefore $\frac{1}{2}g_I(g_I + 1) = (2I + 1)(I + 1)$ nuclear spin functions of A symmetry. The ratio of the number of A spin functions to the number of B spin functions is therefore $(g_I + 1)/(g_I - 1) = (I + 1)/I$.

If the two nuclei are fermions, the overall symmetry (including nuclear spin) must be B, hence the $(2I + 1)(I + 1)$ nuclear spin functions of A symmetry must be associated with rovibronic states of symmetry B. Similarly the $(2I + 1)I$ nuclear spin functions of B symmetry can only occur in combination with rovibronic states of A symmetry. On the other hand if the nuclei are bosons then the overall symmetry must be A; A spin functions now go with A rovibronic states and B spin functions with B rovibronic states. In either case, the rovibronic wavefunctions must be classified with a symmetry label of the CNP group.

The second part of the problem is to decide which rovibronic states are A and which are B. This is reasonably straightforward for a diatomic molecule. A set of symmetry operations can be constructed to have the effect of permuting the two nuclei, see Fig. 1.7. The first such operation is to rotate the molecule about $180°$, which interconverts the two nuclei, but also rotates the electronic wavefunction and the vibrational wavefunction. The rotational wavefunction

In the special case where $I = 0$ there are no antisymmetric nuclear spin functions (the only possible value of M_I is 0). Thus all rovibronic states with B symmetry are missing. This is the case for the O_2 molecule and the CO_2 molecule.

(a spherical harmonic, see *Quantum mechanics 1*, Section 3.4) changes sign under this operation if the rotational quantum number J is odd and remains invariant if J is even. The vibrational wavefunction of a diatomic molecule is totally symmetric in the molecular point group and so does not change under any symmetry operation. The electronic wavefunction is totally symmetric for closed shell molecules, but may not be for molecules with unpaired electrons. The C_2 operation is equivalent to an inversion followed by a reflection in a plane containing the bond axis. The symmetry with respect to inversion is given in the electronic term symbol by the g or u subscript, and the reflection symmetry is given by the superscript $+$ or $-$. Most molecules are closed shell, so the electronic symmetry is Σ_g^+, which is invariant to both operations. For a closed shell molecule we conclude that odd rotational levels are B and even rotational levels are A. Thus, for example, for the H_2 molecule ($I = \frac{1}{2}$) the odd rotational levels are associated with the symmetric nuclear spin functions, and have a nuclear spin degeneracy of 3, whereas the even rotational levels go with the antisymmetric nuclear spin function which has degeneracy 1. The rotational Raman spectrum of H_2, represented in Fig. 1.8, shows a 3:1 intensity alternation superimposed on the usual pattern associated with thermal populations. This alternation is entirely due to the alternation in nuclear spin degeneracy.

In the N_2 molecule ($I = 1$) the odd rotational states have a nuclear spin degeneracy of 6 and the even levels of 12. The intensity alternation in the Raman spectrum is 2:1, and is the other way round from the H_2 molecule, that is, the even levels are more densely populated.

The O_2 molecule ($I = 0$) is an unusual case of a molecule with an open shell configuration in the ground state. The electronic symmetry is Σ_g^-, so the odd rotational levels are A, whereas the even levels are B. Since only the A nuclear spin combination is possible we conclude that all the even rotational

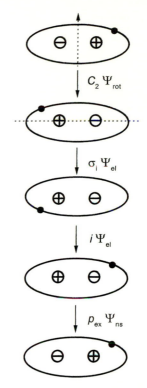

Fig. 1.7 A set of symmetry operations which has the effect of permuting the two nuclei. The 180° rotation operates on the whole rovibronic wavefunction. This is followed by a reflection and an inversion which operate on the electronic wavefunction for a diatomic molecule. (For a linear molecule the vibrational symmetry must also be considered.)

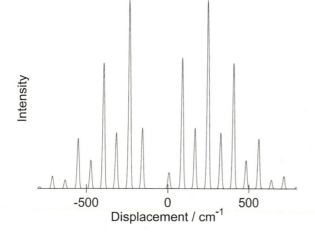

Fig. 1.8 The rotational Raman spectrum of H_2. The Rayleigh line has been removed.

levels in O_2 are missing, thus half the lines are completely missing from the Raman spectrum.

Polyatomic molecules

Similar analyses can be made for polyatomic molecules, where we also find alternating nuclear spin degeneracies in the possible rotational energy levels. The proper analysis requires us to be able to work out the symmetries of the nuclear spin functions and of the rovibronic wavefunction in the permutation group. Only those combinations which finally give the alternating irrep for fermions, or the totally symmetric irrep for bosons are possible.

As an example we consider the situation in the ammonia molecule, which contains three H atoms. We have already worked out the correct symmetries for the possible nuclear spin states in S_3 in the discussion of the electronic symmetries of the Li atom above. The combinations $\alpha\alpha\alpha$, and $\beta\beta\beta$ are totally symmetric (A_1). The three functions $\alpha\alpha\beta$, $\alpha\beta\alpha$, and $\beta\alpha\alpha$ generate $A_1 + E$, as does the similar set of functions with $M_I = -\frac{1}{2}$.

We therefore have a total of four states with symmetry A_1 (twelve, including the threefold degeneracy of the nuclear spin on N) and two (six) doubly degenerate states with symmetry E. The A_1 nuclear spin functions can only be combined with A_2 rovibronic wavefunctions to give the correct alternating irrep for the total wavefunction. These rovibronic states will have a nuclear spin degeneracy of 12. Similarly the only direct product of E which gives A_2 is $E \times E$. E rovibronic states therefore have a nuclear spin degeneracy of 6. Finally all rovibronic states that are totally symmetric with respect to the permutations of S_3 must be missing, since there are no nuclear spin functions of A_2 symmetry.

The analysis of the symmetry of the rovibronic wavefunction involves a consideration of the electronic symmetry (nearly all ground state molecules are closed shell and therefore totally symmetric), the vibrational symmetry, and the rotational wavefunction. The detailed consideration of rotational wavefunctions, even for symmetric tops such as ammonia, is beyond the scope of this book. However, even without this analysis we can expect to see 2:1 intensity alternation effects in any spectrum which resolves the individual rotational states. The interested reader who wishes to pursue this further should consult more advanced texts, such as Landau and Lifshitz (1977) or Bunker (1979).

References

Bunker, P. R. (1979). *Molecular symmetry and spectroscopy*. Academic Press, New York.

Green, N. J. B. (1997). *Quantum mechanics 1*. Oxford Chemistry Primer, Oxford.

Landau, L. D. and Lifshitz, E. M., (1977). *Quantum mechanics*, 3rd edn. Pergamon Press, Oxford.

Szabo, A. and Ostlund, N. S. (1982). *Modern quantum chemistry*. Macmillan, New York.

2 Optimization

2.1 The variational method

Although some quantum-mechanical problems can be solved exactly and explicitly, most cannot, even with the aid of symmetry. Instead, it is necessary to find an approximate solution. Approximations are of two basic types:

- Simplification to a form that permits explicit solution, such as the approximation of a molecular vibration by a harmonic oscillator or a Morse oscillator;
- Building an accurate computational solution by superposing large numbers of simple wavefunctions.

Approximation to the ground state of a system systematically overestimates the energy. The best approximation is that which minimizes the energy.

The former type of approximation is extremely useful for gaining physical insight into the behaviour of the system, whereas the latter type is used for accurate computations.

Whatever approximation is used, it must be optimized. That is, we need a foolproof way to find the best possible approximation of the assumed form. The most suitable and widely used optimization method is the *variational method*, which we have already encountered in *Quantum mechanics 1*.

The basis of the variational method, proved in *Quantum mechanics 1*, Section 1.7, is that the expectation energy calculated from an approximate wavefunction must be greater than the true ground state energy. If we attempt to describe a system in its ground state with an approximate wavefunction containing a variable parameter, then the best wavefunction of this form will be the one that gives the lowest expectation energy, because this will be the closest to the true energy. The variational method consists of finding the value of the parameter that minimizes the energy.

2.2 The hydrogen-like atom

Consider a hydrogen-like atom with nuclear charge Z in its ground state, that is, with one electron in a $1s$ orbital. In atomic units (see *Quantum mechanics 1*, p. 27) the Schrödinger equation for the electronic wavefunction is

$$-\frac{1}{2}\nabla^2\psi - \frac{Z}{r}\psi = E\psi \tag{2.1}$$

The ground-state wavefunction for this system is given in *Quantum mechanics 1*, Chapter 3; nonetheless, it is instructive to look for the best possible solution of the form $\psi = N\exp(-\alpha r)$, which is of the same form as the $1s$ orbital. In this case the variational method, with α as the parameter to be varied, should give the exact solution.

First we find N from the normalization integral:

$$1 = N^2 \int_0^\infty 4\pi r^2 \exp(-2\alpha r)\, dr = \frac{N^2 \pi}{\alpha^3} \tag{2.2}$$

so that $N = (\alpha^3/\pi)^{1/2}$. The integral can be evaluated in many different ways: by parts, by differentiation under the integral sign, by residues, and so on. In this example we shall use several integrals of this form and so we quote the standard result:

$$\int_0^\infty x^n \exp(-sx)\, dx = \frac{n!}{s^{n+1}} \tag{2.3}$$

The integral we have just used is the case $n = 2$, $s = 2\alpha$.

In order to evaluate the expectation energy we need to perform the integral $< \psi|\hat{H}|\psi >$. Applying the Hamiltonian to ψ, we find

$$\hat{H}Ne^{-\alpha r} = -\frac{1}{2}\left[\alpha^2 - \frac{2\alpha}{r}\right]Ne^{-\alpha r} - \frac{Z}{r}Ne^{-\alpha r} \tag{2.4}$$

so that the expectation energy is given by

$$< E > = N^2 \int_0^\infty 4\pi r^2 e^{-\alpha r} \hat{H} e^{-\alpha r}\, dr$$

$$= 4\pi N^2\left(-\left[\frac{1}{8\alpha} - \frac{1}{4\alpha}\right] - \frac{Z}{4\alpha^2}\right)$$

$$= \frac{\alpha^2}{2} - Z\alpha \tag{2.5}$$

This function is a quadratic equation of the variable α and it is elementary to find that it passes through a minimum when $\alpha = Z$. Substituting this value of α back into the expression for the expectation energy gives an energy of $-Z^2/2$, which is the exact ground state energy of the hydrogen-like atom in atomic units (apart from a small correction for the finite nuclear mass).

Problem 2.2.1. *Repeat this analysis for the lowest energy wavefunction with the symmetry of a p_z orbital,* $\psi = Nr\,cos\theta\,exp(-\alpha r).$

2.3 The helium atom

As a less trivial example consider the ground state of a helium-like atom, that is, an atom with two electrons and nuclear charge Z. In the orbital approximation the helium atom has the ground electron configuration $1s^2$ and the orbital function $\psi_{1s}(\mathbf{r}_1)\psi_{1s}(\mathbf{r}_2)$, which we abbreviate $|1s_1 1s_2 >$. This orbital function is symmetric with respect to permutation of the electrons and is associated with an antisymmetric singlet spin function.

However, in the He atom there are not only attractive interactions between each electron and the nucleus, there is also an electron–electron repulsion. The Schrödinger equation in atomic coordinates is therefore

$$-\frac{1}{2}\nabla_1^2\psi - \frac{Z}{r_1}\psi - \frac{1}{2}\nabla_2^2\psi - \frac{Z}{r_2}\psi + \frac{1}{r_{12}}\psi = E\psi \qquad (2.6)$$

The Hamiltonian is the sum of two one-electron hydrogen-like Hamiltonians, one for each electron, and the electron–electron repulsion term.

$$\hat{h}_1\psi + \hat{h}_2\psi + \frac{1}{r_{12}}\psi = E\psi \qquad (2.7)$$

The electron–electron repulsion correlates the motion of the two electrons and ensures that the orbital approximation $|1s_1 1s_2 >$ is not an exact description of the electronic wavefunction of the atom. In this example we use the variational method to find the best possible description of the atom of the assumed form.

We guess that the $1s$ wavefunction has same form as for the H atom, discussed in the previous example, $N\exp(-\alpha r)$, and minimize the expectation energy with respect to the parameter α. The expectation energy is the sum of three terms,

$$
\begin{aligned}
< E > &= < 1s_1 1s_2|\hat{h}_1|1s_1 1s_2 > \\
&\quad + < 1s_1 1s_2|\hat{h}_2|1s_1 1s_2 > + < 1s_1 1s_2|\frac{1}{r_{12}}|1s_1 1s_2 > \\
&= < 1s_1|\hat{h}_1|1s_1 > + < 1s_2|\hat{h}_2|1s_2 > \\
&\quad + < 1s_1 1s_2|\frac{1}{r_{12}}|1s_1 1s_2 > \qquad (2.8)
\end{aligned}
$$

Problem 2.3.1. *Verify that the third term is given by eqn 2.9.*

The simplification occurs because $\hat{h}_1$ only depends on the coordinates of electron 1 and $|1s >$ is normalized. We have already evaluated the first two terms in the preceding example, they are both given by eqn 2.5. The third term is more difficult, but it can be found by rotating the coordinates of $\mathbf{r}_2$ so that the z_2 axis is parallel to $\mathbf{r}_1$. r_{12} can then be expressed in terms of r_1, r_2, and θ_2, using the cosine rule. The integral is then elementary, and gives the result

$$< 1s_1 1s_2|\frac{1}{r_{12}}|1s_1 1s_2 >= \frac{5\alpha}{8} \qquad (2.9)$$

Combining all three terms we find the expectation energy to be

$$< E >= \alpha^2 - 2\alpha\left(Z - \frac{5}{16}\right) \qquad (2.10)$$

which is plotted as a function of α in Fig. 2.1. If we simply used the hydrogen-like value of the parameter $\alpha = Z$, we would obtain the expectation energy $-Z(Z - \frac{5}{8})$, which is -2.75 Hartree for He. However, the minimum expectation energy is obtained when $\alpha = Z - \frac{5}{16}$, giving an energy of $-(Z - \frac{5}{16})^2$, which is -2.85 Hartree for He. The exact ground state energy of He is -2.90 Hartree, so the variational approximation is much better than the use of unmodified $1s$ orbitals. The modification of the optimum value of α below the value of the nuclear charge can be thought of as a reduction of the effective nuclear charge.

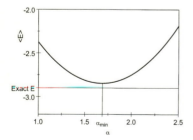

Fig. 2.1 Dependence of the expectation energy for the He atom on the parameter α

This result shows that the electron–electron repulsion reduces the effective nuclear charge felt by the electrons. The best description of the form $|1s_1 1s_2 >$ is where the α parameters are characteristic of a nuclear charge $Z - \frac{5}{16}$. The nuclear charge felt by each electron is said to be partly *screened* by the repulsion of the other electron.

2.4 Optimizing superpositions

In most quantitative applications of quantum chemistry, wavefunctions are constructed by forming a superposition, or linear combination, of many simpler wavefunctions. The most familiar example is the construction of molecular orbitals as linear combinations of atomic orbitals, the LCAO method. The variational method is used to construct the best possible superposition by minimizing the ground state energy with respect to the coefficients.

Thus we want to approximate the true wavefunction by a linear combination of the form $\psi = \sum_j c_j |j >$, where the $|j >$ are some given set of functions. We therefore need to find the coefficients c_j which minimize the expectation energy,

$$< E >=< \psi|\hat{H}|\psi >= \sum_{j,k} c_k c_j < k|\hat{H}|j > \qquad (2.11)$$

However, the coefficients c_j are not all independent. They are constrained by the necessity for the wavefunction to be normalized,

$$< \psi|\psi >= \sum_{j,k} c_k c_j < k|j >= 1 \qquad (2.12)$$

In the combination of atomic orbitals to form a molecular orbital the constituent atomic orbitals on different atoms are not in general orthogonal, so it is not possible to simplify the normalization summation to $\sum_j c_j^2$. The quantity $< k|j >$ is called the *overlap integral* of the constituent functions for obvious reasons, and in matrix notation it is written S_{kj}. The corresponding Hamiltonian matrix element is denoted H_{kj}.

The mathematical method used to minimize a function subject to a constraint is the method of *Lagrange multipliers*, which is described in many textbooks of mathematics for scientists (Stephenson 1973). Differentiating the function $< E >$ and the constraint with respect to one of the coefficients, c_i, and combining the derivatives with a Lagrange multiplier λ we find

$$\sum_j H_{ij} c_j - \lambda \sum_j S_{ij} c_j = 0 \qquad (2.13)$$

This is a linear equation relating the coefficients c_j. A different equation is obtained by differentiating with respect to each possible coefficient c_i. Simultaneous equations obtained in this way are called *secular equations*.

Multiplying each equation by the c_i used to derive it and summing over all c_i we find

$$\sum_{ij} H_{ij} c_i c_j - \lambda \sum_{ij} S_{ij} c_i c_j = 0 \qquad (2.14)$$

A superposition of wavefunctions can be optimized by setting up and solving a set of secular equations.

But the first summation is the expectation energy $< E >$ and the second summation is equal to 1, by eqns 2.11 and 2.12. Hence the Lagrange multiplier λ is equal to $< E >$.

If the coefficients are combined to form the vector $\mathbf{c}$ then the secular equations can be written in the compact form:

$$(\mathbf{H} - < E > \mathbf{S})\mathbf{c} = \mathbf{0} \tag{2.15}$$

where $\mathbf{0}$ is the zero vector.

If the matrix $(\mathbf{H} - < E > \mathbf{S})$ has an inverse, the only possible solution of the secular equations is $\mathbf{c} = \mathbf{0}$, which is not what we are looking for. For a non-trivial solution to exist the matrix must be singular, that is, its determinant, the *secular determinant*, must be zero

$$|\mathbf{H} - < E > \mathbf{S}| = 0 \tag{2.16}$$

On expanding the secular determinant, we obtain a polynomial for $< E >$. If there are n coefficients in the superposition then in general we will obtain an nth order polynomial, which has n roots. Each root is interpreted as the energy of a molecular orbital, and can then be substituted back into the secular equations to obtain the set of coefficients that determines the orbital.

We have derived the secular equations by attempting to minimize the lowest energy orbital, and in return the secular equations have given us expressions for energies of n different orbitals. It can be shown that every orbital energy obtained in this way is greater than the true orbital energy (Levine 1983), so we have variational bounds on all n energy levels.

2.5   The hydrogen molecule ion

The preceding discussion is rather abstract, so we illustrate the ideas by reference to the simplest conceivable system with a chemical bond, the H_2^+ ion. In atomic units the electronic wavefunction obeys the Schrödinger equation

$$-\frac{1}{2}\nabla^2\psi - \frac{1}{r_a}\psi - \frac{1}{r_b}\psi + \frac{1}{R}\psi = E\psi. \tag{2.17}$$

where r_a and r_b are the distances from the electron to nucleus a and nucleus b, respectively, and R is the internuclear distance. Although this equation can be solved exactly by introducing a suitable set of coordinates (Pauling and Wilson 1985), we wish to describe the bonding using the LCAO approximation, in which the valence orbitals on the constituent atoms are superposed to form the molecular orbital. The valence orbitals of the H atoms are the $1s$ orbitals centred on the two nuclei, which we denote $|1 >$ and $|2 >$. We are therefore searching for the best possible wavefunction of the form $c_1|1 > + c_2|2 >$.

The analysis of the preceding section leads to the secular equations

$$(H_{11} - ES_{11})c_1 + (H_{12} - ES_{12})c_2 = 0$$
$$(H_{21} - ES_{21})c_1 + (H_{22} - ES_{22})c_2 = 0 \tag{2.18}$$

We can immediately see, by symmetry, that $H_{11} = H_{22}$, and we denote this quantity α; from normalization $S_{11} = S_{22} = 1$. Furthermore, because the $1s$ orbitals are real, we have $S_{12} = S_{21}$, denoted S, and $H_{12} = H_{21}$, which we denote β.

- α is the expectation energy of an electron in a $1s$ orbital in the presence of both nuclei.
- β is measures the extent to which the orbital energies are altered by the interference of the two orbitals.
- S is the *overlap integral* of the two orbitals.

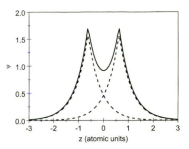

Fig. 2.2 A cut through the optimized LCAO wavefunction for the H_2^+ molecular orbital along the z axis.

The secular equations can now be rewritten

$$(\alpha - E)c_1 + (\beta - ES)c_2 = 0$$
$$(\beta - ES)c_1 + (\alpha - E)c_2 = 0 \qquad (2.19)$$

The unknowns in the secular equations are the energy E and the constants c_1 and c_2, as the quantities α, β, and S can all be evaluated from the $1s$ orbitals by integration.

For these simultaneous equations to have a non-trivial solution the secular determinant must be zero:

$$\begin{vmatrix} \alpha - E & \beta - ES \\ \beta - ES & \alpha - E \end{vmatrix} = (\alpha - E)^2 - (\beta - ES)^2 = 0 \qquad (2.20)$$

which implies

$$E = \frac{\alpha \pm \beta}{1 \pm S} \qquad (2.21)$$

Substituting the energy $(\alpha + \beta)/(1 + S)$ back into the secular equations, we find that $c_1 = c_2$, giving the molecular orbital, after normalization,

$$\psi_g = \frac{|1> + |2>}{\sqrt{2(1 + S)}} \qquad (2.22)$$

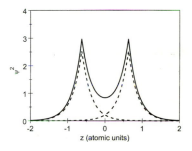

Fig. 2.3 A cut through the optimized LCAO electron density for the H_2^+ molecular orbital along the z axis.

The wavefunction and the corresponding probability density are shown in Figs 2.2 and 2.3. This superposition of $|1>$ and $|2>$ results in constructive interference between the two AOs in the region of overlap. The molecular orbital is thus a bonding orbital, and its energy is reduced relative to α (β is negative).

Similarly, for the energy $(\alpha - \beta)/(1 - S)$, we find the orbital

$$\psi_u = \frac{|1> - |2>}{\sqrt{2(1 - S)}} \qquad (2.23)$$

This wavefunction is antibonding, as can be seen from Fig. 2.4. The superposition results in destructive interference in the region of overlap, thus removing electron density from the bonding region. Its energy is increased relative to α.

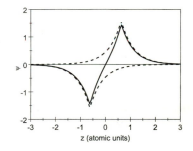

Fig. 2.4 A cut through the optimized LCAO wavefunction for the H_2^+ antibonding orbital along the z axis.

We could, instead, have used the symmetry arguments of *Quantum mechanics 1*, Chapter 4 to show that the only possible symmetry-adapted linear combinations are those where $c_1 = \pm c_2$. The variational method clearly arrives at the same result.

The problem is now solved apart from the evaluation of the integrals, which can be evaluated explicitly. The mathematics is elementary but tedious, and we simply outline the method and the results.

First we consider α and β. The Hamiltonian, eqn 2.17, splits into three terms: the Hamiltonian for hydrogen atom a, the potential energy of repulsion between the two nuclei (which is constant if the internuclear distance is fixed), and the potential energy of attraction between the electron and nucleus b. The energy of the ground state H atom in atomic units is $-\frac{1}{2}$, therefore

$$\alpha = -\frac{1}{2} + \frac{1}{R} - < 1|\frac{1}{r_b}|1 > \tag{2.24}$$

$$\beta = (-\frac{1}{2} + \frac{1}{R})S - < 2|\frac{1}{r_b}|1 > \tag{2.25}$$

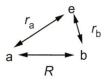

The remaining integrals, $< 1|\frac{1}{r_b}|1 >$, $< 2|\frac{1}{r_b}|1 >$ and $< 2|1 >$, are called the *Coulomb integral*, the *resonance integral*, and the *overlap integral* respectively. They are most easily evaluated by using elliptical coordinates μ, ν, and ϕ for the electron, defined by

Fig. 2.5 Variables for elliptical coordinates.

$$\mu = \frac{r_a + r_b}{R} \tag{2.26}$$

$$\nu = \frac{r_a - r_b}{R} \tag{2.27}$$

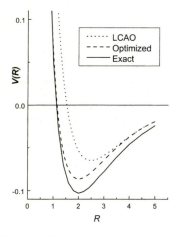

(see Fig. 2.5). If the two nuclei lie on the z-axis, then with the electron they define a plane which includes the z-axis. The angle ϕ is the angle between this plane and the xz plane. The volume element in elliptical coordinates (the Jacobian) is given by

$$dx\ dy\ dz = \frac{R^3}{8}(\mu^2 - \nu^2)d\mu\ d\nu\ d\phi \tag{2.28}$$

All space is covered by permitting μ to vary from 1 to ∞, ν from -1 to 1, and ϕ from 0 to 2π.

Substituting the explicit forms of $|1 >$ and $|2 >$, transforming to elliptical coordinates and integrating, we find

$$< 1|2 > = e^{-R}(1 + R + \frac{1}{3}R^2) \tag{2.29}$$

$$< 1|\frac{1}{r_2}|1 > = \frac{1}{R}[1 - e^{-2R}(1 + R)] \tag{2.30}$$

$$< 2|\frac{1}{r_2}|1 > = e^{-R}(1 + R) \tag{2.31}$$

Fig. 2.6 Calculated potential energy curves for the H$_2^+$ ion.

Problem 2.5.1. *Perform the integrations leading to eqns 2.29 to 2.31.*

When these are inserted into the expressions for α and β, they give an explicit form for the potential energy function, shown in Fig. 2.6. Comparison with the exact solution, also shown in Fig. 2.6, shows that the simple LCAO approximation is not an accurate description of the molecular orbital.

Several obvious improvements can be made to this bonding model. We can introduce a variational parameter into the constituent $1s$ orbitals, as we did for the helium atom. The idea behind this change is to allow for the fact that the electron is attracted to both nuclei; for example, if the nuclei are close together the constituent $1s$ orbitals might be expected to contract. The effect of this refinement is also shown in Fig. 2.6. It can be seen to be a significant improvement.

A second improvement would be to superpose combinations of higher energy atomic orbitals with the correct symmetry. For example the ψ_g orbital, which has symmetry σ_g, can also contain contributions from $2s_1 + 2s_2$ and from $2p_{z1} - 2p_{z2}$. This strategy of expanding the basis set is a common strategy for refining the description of bonding in more complicated molecules, because it can be implemented relatively easily on a computer.

2.6 The hydrogen molecule – configuration interaction

The discussion in the preceding section deals with a particularly simple molecule, because it contains only one electron. Most molecules of interest in chemistry are bonded by electron pairs. In this section we therefore consider how the LCAO method can be extended to the hydrogen molecule.

The hydrogen molecule contains two electrons and so, in addition to the electrostatic attractions between each electron and the two nuclei and the repulsion between the two nuclei, there is also an electrostatic repulsion between the two electrons. Furthermore, as discussed in Chapter 1, the electronic wavefunction must be antisymmetric with respect to permutation of the two electrons.

The electronic Hamiltonian for the hydrogen molecule in the Born–Oppenheimer approximation is

$$\hat{H} = \frac{1}{R} + \left[-\frac{1}{2}\nabla_1^2 - \frac{1}{r_{1a}} - \frac{1}{r_{1b}} \right]$$
$$+ \left[-\frac{1}{2}\nabla_2^2 - \frac{1}{r_{2a}} - \frac{1}{r_{2b}} \right] + \frac{1}{r_{12}} \qquad (2.32)$$

where a and b denote the two nuclei and 1 and 2 the two electrons. This Hamiltonian is the sum of four terms. The first term, the inter-nuclear repulsion energy, is independent of the coordinates of the electrons. The second term represents the kinetic energy of electron 1 and the potential energy of its interaction with the two nuclei – it depends on the coordinates of electron 1 only; likewise the third term depends on the coordinates of electron 2 only. These terms are known as single-electron operators. The final term, the electron–electron repulsion, is a two-electron operator, because it depends on the coordinates of both electrons. We therefore abbreviate the Hamiltonian in the form

$$\hat{H} = \frac{1}{R} + \hat{h}_1 + \hat{h}_2 + \frac{1}{r_{12}} \qquad (2.33)$$

The electronic Hamiltonian of any molecule is essentially similar to this. It contains a constant nuclear repulsion energy, a single-electron operator for each electron, representing the electron kinetic energy and the electron-nucleus attractions, and a two-electron operator for each possible combination of two electrons, representing the electron-electron repulsions.

In the simplest approximation, the *minimal basis* description, the molecular orbitals are the same as those for H_2^+, $(|1> + |2>)/\sqrt{2(1+S)}$ for the Σ_g^+ bonding orbital ($|g>$), and $(|1> - |2>)/\sqrt{2(1-S)}$ for the antibonding orbital ($|u>$).

We expect that both electrons will occupy the bonding orbital in the ground electron configuration. Let us denote the resulting product wavefunction $|gg>$. To ensure the correct symmetry with respect to permutation of the electrons this orbital wavefunction must be multiplied by an antisymmetric, singlet, spin function $(\alpha\beta - \beta\alpha)/\sqrt{2}$ (see Chapter 1).

The expectation energy is given by

$$E = 2 <g|\hat{h}_1|g> + <gg|\frac{1}{r_{12}}|gg> \tag{2.34}$$

where we have used the normalization property $<g|g> = 1$ and the fact that $<g|\hat{h}_1|g> = <g|\hat{h}_2|g>$, since the single-electron operators differ only by the label on the electron.

Let us now consider all the possible electron configurations in this minimal basis description of the hydrogen molecule. The configuration of lowest energy is where both electrons occupy the σ_g bonding orbital. This configuration has the term symbol $^1\Sigma_g^+$ (see *Quantum mechanics 1*, Chapter 4 and Chapter 1). The first excited configuration has one electron excited to the σ_u orbital, giving terms $^1\Sigma_u^+$ and $^3\Sigma_u^+$, whereas the second excited configuration has both electrons in the antibonding orbital, and gives a second $^1\Sigma_g^+$ term.

Both the ground configuration and the doubly excited configuration give wavefunctions with the same symmetry, so it is natural to ask whether we can obtain an improved description of the ground state wavefunction by mixing them, that is, by finding coefficients c_1 and c_2 that minimize the energy of the combination $c_1|gg> + c_2|uu>$. This method, in which we improve the ground-state wavefunction by mixing in excited configurations with the same symmetry, is known as *configuration interaction*, or CI for short, and is one of the most important methods in modern quantum chemistry.

Since $|gg>$ and $|uu>$ are orthogonal, the overlap integral is zero, and the secular equations can be written

$$(<gg|\hat{H}|gg> -E)c_1 + <gg|\hat{H}|uu> = 0$$
$$<uu|\hat{H}|gg> c_1 + (<uu|\hat{H}|uu> -E)c_2 = 0 \tag{2.35}$$

By splitting the Hamiltonian into one-electron and two-electron terms the matrix elements can be simplified to

$$<gg|\hat{H}|gg> = 2 <g|\hat{h}_1|g> + <gg|r^{-1}|gg>$$
$$<gg|\hat{H}|uu> = <uu|\hat{H}|gg> = <uu|r^{-1}|gg> \tag{2.36}$$
$$<uu|\hat{H}|uu> = 2 <u|\hat{h}_1|u> + <uu|r^{-1}|uu>$$

The single-electron integrals $<g|\hat{h}_1|g>$ and $<u|\hat{h}_1|u>$ are the energies of the $|g>$ and $|u>$ orbitals (the same as in the H_2^+ ion), and are denoted ϵ.

Two-electron integrals in which each electron occupies the same orbital in both the bra and the ket are known as *Coulomb integrals*, and are denoted J. Two-electron integrals where the electrons occupy different orbitals in the bra and the ket are known as *exchange integrals*, and denoted K. Thus we have

$$J_{gg} = <gg|r^{-1}|gg>$$

$$J_{gu} = <gu|r^{-1}|gu> = <ug|r^{-1}|ug>$$

$$J_{uu} = <uu|r^{-1}|uu>$$

$$K_{gu} = <uu|r^{-1}|gg> = <gg|r^{-1}|uu>$$

$$= <gu|r^{-1}|ug> = <ug|r^{-1}|gu> \qquad (2.37)$$

Problem 2.6.1. *Show that the four expressions for K_{gu} are identical.*

The ground state energy is therefore the lowest eigenvalue of the CI matrix

$$\begin{pmatrix} 2\epsilon_g + J_{gg} & K_{gu} \\ K_{gu} & 2\epsilon_u + J_{uu} \end{pmatrix}$$

Because the off-diagonal elements of this matrix are non-zero it is clear that further improvement of the variational energy can be obtained by the CI method. This improvement is notable when the molecule is close to dissociation. The wavefunction for the ground configuration can be expanded in terms of the constituent $1s$ orbitals,

$$|gg> = \frac{|11> + |12> + |21> + |22>}{2(1+S)} \qquad (2.38)$$

When the molecule is close to dissociation this wavefunction is not a good description of the molecule because it gives a probability of $\frac{1}{2}$ to ionic configurations ($|11>$ and $|22>$), in which both electrons are found in the same $1s$ orbital, and are therefore close to the same nucleus. However, H_2 always dissociates into two H atoms, rather than giving ions half the time. For this reason Heitler and London proposed an alternative to the molecular orbital description, which is more appropriate for large inter-nuclear distances, known as *valence bond* theory.

The valence bond orbital wavefunction is given by

$$\psi_{VB} = \frac{|12> + |21>}{\sqrt{2(1+S^2)}} \qquad (2.39)$$

neglecting the ionic contributions. The valence bond description is good close to dissociation, but needs to have the ionic terms $|11>$ and $|22>$ mixed in at smaller internuclear separations.

The minimal basis CI description leads to the best wavefunction of the form

$$c_1|gg> + c_2|uu> = \frac{1}{2}(c_1 + c_2)(|11> + |22>)$$

$$+ \frac{1}{2}(c_1 - c_2)(|12> + |21>) \qquad (2.40)$$

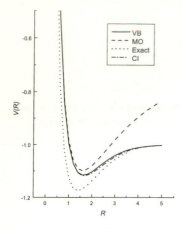

Fig. 2.7 H_2 molecule PE curves.

which can be thought of either as the best possible combination of electron configurations in the molecular orbital theory, or as the best possible combination of the valence bond wavefunction with ionic terms, or as the best possible combination of simple MO theory and simple VB theory. For example, the inclusion of CI as a refinement of the molecular orbital theory leads to the correct dissociation products. Simple MO, VB, and CI potential energy curves for H_2 are shown in Fig. 2.7.

The molecular orbital and valence bond theories are both approximations to reality. Both theories can be refined; molecular orbital theory by CI and VB theory by the inclusion of ionic terms. These refinements lead essentially to the same picture. For more details see McWeeny (1992).

Lowest triplet state of H_2

We complete our discussion of hydrogen by considering the lowest triplet state. The triplet spin functions, such as $\alpha\alpha$, are symmetric with respect to permuting the electrons, so the orbital function must be antisymmetric. The lowest configuration that can give an antisymmetric orbital wavefunction is where one electron is excited into the σ_u antibonding orbital, giving a $^3\Sigma_u^+$ term. There is also a singlet state from the same configuration, giving a $^1\Sigma_u^+$ term. We shall consider both of these together, since both energy levels arise from the same electron configuration. The resulting orbital functions are

$$\frac{1}{\sqrt{2}}(|gu> \mp |ug>) \tag{2.41}$$

where the $-$ sign refers to the triplet state and the $+$ sign the singlet. The expectation energy is found by the usual method of splitting the Hamiltonian into one- and two-electron terms, with the result

$$E_\mp = \epsilon_g + \epsilon_u + J_{gu} \mp K_{gu} \tag{2.42}$$

The singlet and triplet wavefunctions have different energies as a result of the exchange correction K_{gu}, which comes from the antisymmetrization of the electronic wavefunction with respect to permuting the two electrons. Had we simply used the product wavefunction $|gu>$, which does not recognize this permutation symmetry, the exchange energy would not have appeared.

Nearly all closed shell molecules have excited triplet states, whose energies are lower than the excited singlet states with the same electron configuration, because the electron repulsion in the antisymmetrized triplet state is less than that in the orbitally symmetric singlet state, as discussed in Chapter 1.

References

Green, N. J. B. (1997). *Quantum mechanics 1*. Oxford Chemistry Primer, Oxford.

Levine, I. N. (1983). *Quantum chemistry*. Allyn and Bacon, Boston.

McWeeny, R. (1992). *Methods of molecular quantum mechanics*. Academic Press, London.

Pauling, L. and Wilson, E. B. (1985). *Introduction to quantum mechanics*. Dover, New York, Section 42c.

Stephenson, G. (1973). *Mathematical methods for science students*. Longman, London.

3 Perturbations

Perturbation theory is a systematic method for successive refinement of an approximation in the form of a series expansion. To be useful, the series should converge rapidly.

Many of the quantum mechanical problems that we have considered thus far have been solved by making simple approximations. These approximations are frequently successful in giving us a framework for understanding the properties of atoms and molecules, for example the separation of electronic and nuclear coordinates (the Born–Oppenheimer approximation, *Quantum mechanics 1*, Section 2.5) enables us to define a potential energy surface for the motion of the nuclei in a molecule. Similarly, the rigid rotor approximation enables us to treat the rotational and vibrational motions of a molecule separately. These approximations are tremendously useful; they give us insight, which we could not otherwise have gained, but they are approximations. Some important physical and chemical phenomena are not predicted by these simple approximations; for example, the optical $(d - d)$ transitions of an octahedral transition metal complex are forbidden by the electronic selection rule (they originate and terminate in electronic terms with *g* symmetry, and the electric dipole moment operator has *u* symmetry), yet we know that many transition metal complexes are coloured as a result of these 'forbidden' transitions. Another example is the treatment of molecular vibrations, which can be described to a first approximation as harmonic. However, in a harmonic oscillator the selection rule $\Delta v = \pm 1$ must be obeyed. In a real molecule, although the strongest transitions usually obey $\Delta v = \pm 1$, weaker transitions with $\Delta v = \pm 2, 3, \ldots$ (overtones) are also observed, and the energy levels deviate from the equally spaced ladder expected for a harmonic oscillator.

We clearly do not want to throw away the physical insight given by successful approximations, so instead we search for ways of refining them. The method discussed in this chapter is called *perturbation theory*, in which the simple theory is used as a starting point for a series of successive refinements. The difference between the true Hamiltonian and the approximate Hamiltonian is assumed to be a small perturbation of the system, and successive levels of refinement are obtained under the assumption that the effects of the perturbation are small.

There are many other examples of important problems in chemistry that can be treated using perturbation theory. For example, the orbital approximation (*Quantum mechanics 1*, Section 2.7) can be corrected for electron–electron repulsions, and the formation of a chemical bond as two atoms approach one another and perturb each other's electron distributions can be dealt with in this way. The same method can be used to investigate the response of a system to a small perturbation applied from outside, such as an electric or magnetic field. We shall discuss applications to external fields in the next chapter.

3.1   Perturbation theory

We start by introducing the mathematical apparatus of perturbation theory. The basic idea is that the true Hamiltonian $\hat{H}$ is close to an approximate Hamiltonian $\hat{H}^{(0)}$, whose eigenfunctions and eigenvalues are known. The difference, $\hat{H}^{(1)}$, is called a perturbation (higher order corrections, $\hat{H}^{(2)}$ etc. can also be included, if desired). We also introduce a parameter λ to keep track of the *order of refinement*, so that the first-order correction $\hat{H}^{(1)}$ is multiplied by λ, $\hat{H}^{(2)}$ by λ^2 and so on. Thus, the corrected Hamiltonian is given by

$$\hat{H} = \hat{H}^{(0)} + \lambda\hat{H}^{(1)} + \lambda^2\hat{H}^{(2)} + \ldots \tag{3.1}$$

> The parameter λ is a dummy parameter, which simply keeps track of the order of approximation. The perturbation solution is a power series in λ.

The parameter λ does not represent any real physical quantity, it is simply a mathematical technique for keeping track of the order of approximation. In the end, we will only be interested in putting $\lambda = 1$ in the solution, but if the mathematical method works, it should be equally valid if any fraction λ of the true perturbation is applied. Thus functional dependence of the solution on λ can be interpreted as a dependence on the strength or weight of the perturbation, and we can therefore require that the method should work for all values of λ sufficiently small for the solution, which is a power series in λ, to converge.

Let us denote the unknown eigenfunctions of the true Hamiltonian $\hat{H}$ ψ_i, and the corresponding unknown energies E_i. If the perturbation is small then E_i will be close to some eigenvalue of $\hat{H}^{(0)}$, $E_i^{(0)}$, and ψ_i will be close to the corresponding eigenfunction $\psi_i^{(0)}$. The two states $\psi_i^{(0)}$ and ψ_i are correlated in the sense that, if an unperturbed system is initially in state $\psi_i^{(0)}$, and the perturbation is switched on slowly and gradually, the system will end up in state ψ_i (see Chapter 5).

Since the true system is close to the approximate system, we seek to express successive corrections in the form of a power series in λ, in the hope that such a series will converge:

$$\psi_i = \psi_i^{(0)} + \lambda\psi_i^{(1)} + \lambda^2\psi_i^{(2)} + \ldots \tag{3.2}$$

with a similar series for the energy

$$E_i = E_i^{(0)} + \lambda E_i^{(1)} + \lambda^2 E_i^{(2)} + \ldots \tag{3.3}$$

The parameter λ provides a label on the level of refinement; thus, corrections which are linear in λ are first-order corrections and are labelled with the superscript $^{(1)}$, corrections quadratic in λ are second order, and are (hopefully) smaller than the first-order corrections, and so on.

The results of perturbation theory are derived by taking the exact Schrödinger equation $\hat{H}\psi_i = E_i\psi_i$ and substituting the series for $\hat{H}$, ψ_i, and E_i. If we then group all terms with the same power of λ we find

$$\hat{H}^{(0)}\psi_i^{(0)} + \lambda[\hat{H}^{(0)}\psi_i^{(1)} + \hat{H}^{(1)}\psi_i^{(0)}] +$$
$$\lambda^2[\hat{H}^{(0)}\psi_i^{(2)} + \hat{H}^{(1)}\psi_i^{(1)} + \hat{H}^{(2)}\psi_i^{(0)}] + \ldots =$$
$$E_i^{(0)}\psi_i^{(0)} + \lambda[E_i^{(0)}\psi_i^{(1)} + E_i^{(1)}\psi_i^{(0)}] +$$
$$\lambda^2[E_i^{(0)}\psi_i^{(2)} + E_i^{(1)}\psi_i^{(1)} + E_i^{(2)}\psi_i^{(0)}] + \ldots \tag{3.4}$$

Because this solution must be true for all values of the scaling parameter λ, we can equate the coefficients of powers of λ, giving

$$(\hat{H}^{(0)} - E_i^{(0)})\psi_i^{(0)} = 0 \tag{3.5}$$

$$(\hat{H}^{(0)} - E_i^{(0)})\psi_i^{(1)} + (\hat{H}^{(1)} - E_i^{(1)})\psi_i^{(0)} = 0 \tag{3.6}$$

$$(\hat{H}^{(0)} - E_i^{(0)})\psi_i^{(2)} + (\hat{H}^{(1)} - E_i^{(1)})\psi_i^{(1)} + \\ + (\hat{H}^{(2)} - E_i^{(2)})\psi_i^{(0)} = 0 \tag{3.7}$$

$$(\hat{H}^{(0)} - E_i^{(0)})\psi_i^{(3)} + (\hat{H}^{(1)} - E_i^{(1)})\psi_i^{(2)} + \\ + (\hat{H}^{(2)} - E_i^{(2)})\psi_i^{(1)} + (\hat{H}^{(3)} - E_i^{(3)})\psi_i^{(0)} = 0 \tag{3.8}$$

The first of these equations tells us nothing new. It is simply the original 'zero-order' approximation. The other equations permit us to obtain the corrections to this approximation.

Without any loss of generality we can assume that all corrections to the wavefunction are orthogonal to $\psi_i^{(0)}$. We can then make significant progress by multiplying each equation by $\psi_i^{(0)*}$ and integrating. Normally we are only concerned with a single perturbation, so $\hat{H}^{(2)}$ and higher order perturbations are zero, but for completeness we shall not drop these terms at present. We therefore obtain

$$< \psi_i^{(0)}|(\hat{H}^{(0)} - E_i^{(0)})|\psi_i^{(1)} > + < \psi_i^{(0)}|(\hat{H}^{(1)} - E_i^{(1)})|\psi_i^{(0)} > = 0 \tag{3.9}$$

$$< \psi_i^{(0)}|(\hat{H}^{(0)} - E_i^{(0)})|\psi_i^{(2)} > + < \psi_i^{(0)}|(\hat{H}^{(1)} - E_i^{(1)})|\psi_i^{(1)} > + \\ < \psi_i^{(0)}|(\hat{H}^{(2)} - E_i^{(2)})|\psi_i^{(0)} > = 0 \tag{3.10}$$

$$< \psi_i^{(0)}|(\hat{H}^{(0)} - E_i^{(0)})|\psi_i^{(3)} > + < \psi_i^{(0)}|(\hat{H}^{(1)} - E_i^{(1)})|\psi_i^{(2)} > + \\ < \psi_i^{(0)}|(\hat{H}^{(2)} - E_i^{(2)})|\psi_i^{(1)} > + < \psi_i^{(0)}|(\hat{H}^{(3)} - E_i^{(3)})|\psi_i^{(0)} > = 0 \tag{3.11}$$

In each case the first term is zero, because $\psi_i^{(0)}$ is an eigenfunction of the Hermitian operator $\hat{H}^{(0)}$, and is orthogonal to all the corrections – for example

$$< \psi_i^{(0)}|\hat{H}^{(0)}|\psi_i^{(1)} > = E_i^{(0)} < \psi_i^{(0)}|\psi_i^{(1)} > = 0$$

We therefore find

$$< \psi_i^{(0)}|\hat{H}^{(1)}|\psi_i^{(0)} > = E_i^{(1)} \tag{3.12}$$

$$< \psi_i^{(0)}|\hat{H}^{(1)}|\psi_i^{(1)} > + < \psi_i^{(0)}|\hat{H}^{(2)}|\psi_i^{(0)} > = E_i^{(2)} \tag{3.13}$$

$$< \psi_i^{(0)}|\hat{H}^{(1)}|\psi_i^{(2)} > + < \psi_i^{(0)}|\hat{H}^{(2)}|\psi_i^{(1)} > +$$
$$< \psi_i^{(0)}|\hat{H}^{(3)}|\psi_i^{(0)} > = E_i^{(3)} \qquad (3.14)$$

The first of these tells us that the first-order energy is the expectation of the first-order Hamiltonian, calculated with the zero-order wavefunction. The second tells us that to calculate the second order energy we need to know the first-order corrected wavefunction, and the third equation seems to tell us that to calculate the third-order energy correction we need the second-order wavefunction. This final conclusion is not in fact correct; as Problem 3.1.1 demonstrates, we can actually find $E_i^{(3)}$ directly from the first-order corrected wavefunction.

Problem 3.1.1. *Take eqn 3.6, premultiply by $\psi_i^{(2)*}$ and integrate. Compare the result with eqn 3.7 multiplied by $\psi_i^{(1)*}$ and integrated. Hence eliminate $< \psi_i^{(0)}|\hat{H}^{(1)}|\psi_i^{(2)} >$ from eqn 3.14, and find*

$$< \psi_i^{(1)}|\hat{H}^{(1)} - E_i^{(1)}|\psi_i^{(1)} > + < \psi_i^{(0)}|\hat{H}^{(2)}|\psi_i^{(1)} > +$$
$$< \psi_i^{(1)}|\hat{H}^{(2)}|\psi_i^{(0)} > + < \psi_i^{(0)}|\hat{H}^{(3)}|\psi_i^{(0)} > = E_i^{(3)} \qquad (3.15)$$

If there is only a first-order perturbation only the first term is non-zero.

This is an example of Wigner's theorem, which states that if the wavefunction is known to order n, then the energy can be calculated to order $2n + 1$. We conclude, therefore, that if we can find the first-order corrected wavefunction, then corrections to the energy up to the third order follow immediately.

3.2 The Rayleigh–Ritz method

The theory outlined so far has been at a general level. To use these results we need a systematic method for constructing successive corrections to the wavefunction. The usual approach is to use the method of expansion in eigenfunctions, described in *Quantum mechanics 1*, Section 1.6, and the most convenient orthogonal basis for this purpose is the set of functions $\{\psi_i^{(0)}\}$. (Since $\hat{H}^{(0)}$ is Hermitian, its eigenfunctions must be orthogonal.) We therefore denote $\psi_i^{(0)}$ by $|i>$, and write $\psi_i^{(1)}$ as the expansion

The Rayleigh–Ritz method constructs the perturbation series as a superposition of the eigenfunctions of the zero order Hamiltonian, which are known.

$$\psi_i^{(1)} = \sum_j c_{ij}|j> \qquad (3.16)$$

Finding the first-order correction to the wavefunction is now simply a matter of finding the coefficients c_{ij} in the expansion. Eqn 3.6 for the first order correction now becomes

$$\sum_j c_{ij}\hat{H}^{(0)}|j> + \hat{H}^{(1)}|i> = E_i^{(0)} \sum_j c_{ij}|j> + E_i^{(1)}|i> \qquad (3.17)$$

We can simplify further because $|j>$ is an eigenfunction of $\hat{H}^{(0)}$ with eigenvalue $E_j^{(0)}$, which helps with the first term on the left hand side.

$$\sum_j E_j^{(0)} c_{ij} |j> + \hat{H}^{(1)} |i> = E_i^{(0)} \sum_j c_{ij} |j> + E_i^{(1)} |i> \qquad (3.18)$$

Although this equation looks even worse than the original equation, we are actually now in a position to find the first-order corrections.

The first-order energy is given by eqn 3.12, which in this notation is

$$E_i^{(1)} = <i|\hat{H}^{(1)}|i> \qquad (3.19)$$

To find the correction to the wavefunction we multiply eqn 3.18 by $<k|$ (where $k \neq i$) and integrate.

$$\sum_j E_j^{(0)} c_{ij} <k|j> + <k|\hat{H}^{(1)}|i> =$$

$$E_i^{(0)} \sum_j c_{ij} <k|j> + E_i^{(1)} <k|i> \qquad (3.20)$$

(Remember $|i>$ is the wavefunction whose corrections we are seeking.) Since the zero-order functions are orthogonal and normalized, $<k|j> = 0$ if $k \neq j$, and 1 if $k = j$. Thus, the only non-zero term in each summation is the term with $k = j$, giving

$$E_k^{(0)} c_{ik} + <k|\hat{H}^{(1)}|i> = E_i^{(0)} c_{ik} \qquad (3.21)$$

and hence,

$$c_{ik} = \frac{<k|\hat{H}^{(1)}|i>}{(E_i^{(0)} - E_k^{(0)})} \qquad (3.22)$$

This determines all the coefficients c_{ik} in the summation apart from the coefficient c_{ii}, which represents the component of the correction in the direction of the original approximation. Since the correction is orthogonal to $\psi_i^{(0)}$, c_{ii} must be zero.

The perturbation modifies the wavefunction by superposing other zero-order wavefunctions. Each coefficient in this superposition depends on two factors: the matrix element of the perturbation which mixes the two states, and the energy difference between the states being mixed. Both factors are important – for example, when two atoms are brought together, their orbitals interact and mix, forming molecular orbitals. The extent of the mixing, and therefore the strength of the bond, depends on the overlap of the states and the energy difference.

The rule of thumb that strong bonds require good overlap between the atomic orbitals and good energy matching has its origin in second-order perturbation theory.

Using this expression we can assemble the first-order correction from eqn 3.16.

$$\psi_i^{(1)} = \sum_j{}' \frac{<j|\hat{H}^{(1)}|i> |j>}{(E_i^{(0)} - E_k^{(0)})} \qquad (3.23)$$

The prime $(')$ on the summation indicates that the sum does not include the term with $j = i$.

Problem 3.2.1. *Show from eqns 3.13 and 3.15 that if the Hamiltonian only has a first-order perturbation then the second- and third-order energy corrections are given by*

$$E_i^{(2)} = \sum_j{}' \frac{<i|\hat{H}^{(1)}|j><j|\hat{H}^{(1)}|i>}{E_i^{(0)} - E_j^{(0)}} \qquad (3.24)$$

and

$$E_i^{(3)} = \sum_{jk}{}' \frac{<i|\hat{H}^{(1)}|j><j|\hat{H}^{(1)}|k><k|\hat{H}^{(1)}|i>}{(E_i^{(0)} - E_j^{(0)})(E_i^{(0)} - E_k^{(0)})}$$
$$- <i|\hat{H}^{(1)}|i> \sum_j{}' \frac{<i|\hat{H}^{(1)}|j><j|\hat{H}^{(1)}|i>}{(E_i^{(0)} - E_j^{(0)})^2} \qquad (3.25)$$

where the sums cover all values of $j, k \neq i$.

In order to apply perturbation theory to a particular problem, all we need to know are the unperturbed (zero-order) wavefunction and energy levels; the matrix elements of the perturbation are simply *numbers* that can be calculated from these zero-order eigenfunctions by integration. To illustrate typical applications of perturbation theory we now consider several detailed examples. The first of these has the virtue of being exactly soluble, so that the method can be verified.

Perturbed harmonic oscillator

We illustrate the application of perturbation theory using the example of the effect of an external electric field V on the energy levels of a particle of charge Q oscillating harmonically in one dimension. This problem is a model for the interaction of a harmonic oscillator with a linear external perturbation. It is of central importance in many areas of physics (e.g. quantum electrodynamics, see Feynman and Hibbs 1965) and it has the virtue that it is elementary to solve it exactly, as well as by perturbation theory. The potential energy in the absence of the perturbation is $\frac{1}{2}kx^2$, and the application of the field introduces an extra term VQx into the potential energy. The unperturbed and perturbed potentials are illustrated in Fig. 3.1. To apply perturbation theory we simply need the zero-order energy levels and the matrix elements of the perturbation, $VQ <j|\hat{x}|i>$.

The first-order change in energy level i is given by the expectation of the perturbation operator, $VQ <i|\hat{x}|i>$. This expectation must be identically zero for all values of the quantum number i because all eigenfunctions of the simple harmonic oscillator are either even or odd with respect to inversion through the origin, and x is odd. Thus, regardless of whether $|i>$ is even or odd, the integrand is always odd with respect to inversion, and the integral must be zero.

To find the second-order correction to the energy we need the matrix elements which mix the eigenfunctions. These can be found by direct integration, using the properties of the Hermite polynomials. Alternatively it is

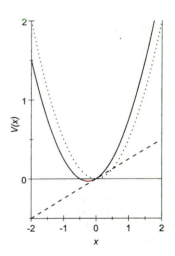

Fig. 3.1 The unperturbed harmonic potential (dotted line), the perturbation dashed line), and the perturbed, but still harmonic potential (full line).

much simpler to use the ladder operators introduced in *Quantum mechanics 1*, Section 3.2.

Ladder operator method.   The ladder operators $R^\pm$ are defined by

$$\hat{R}^\pm = \frac{1}{\sqrt{2\mu}}[\hat{p}_x \pm i\mu\omega\hat{x}] \tag{3.26}$$

where μ is the reduced mass and ω the (angular) frequency of the vibration. The ladder-up operator $\hat{R}^+$ transforms a harmonic oscillator state $|n>$ into $|n+1>$ apart from a constant of proportionality, which we denote c_n^+. Similarly the ladder-down operator reduces the quantum number by one:

$$\hat{R}^\pm |n> = c_n^\pm |n \pm 1> \tag{3.27}$$

These operators are useful for finding the matrix elements of $\hat{x}$ because the operator $\hat{x}$ can be expressed in terms of them,

$$\hat{x} = \frac{1}{i\omega\sqrt{2\mu}}[\hat{R}^+ - \hat{R}^-] \tag{3.28}$$

Hence we can write

$$\hat{x}|n> = \frac{1}{i\omega\sqrt{2\mu}}[\hat{R}^+ - \hat{R}^-]|n>$$
$$= \frac{1}{i\omega\sqrt{2\mu}}[c_n^+ |n+1> - c_n^- |n-1>] \tag{3.29}$$

The only non-zero matrix elements $<j|\hat{x}|n>$ are those for which $j = n \pm 1$, which are given by

$$<n \pm 1|\hat{x}|n> = \frac{\pm 1}{i\omega\sqrt{2\mu}}c_n^\pm \tag{3.30}$$

We therefore need to find the proportionality constants $c_n^\pm$. To do this, we use the fact that the two ladder operators are Hermitian conjugates, so that for example

$$<n-1|\hat{R}^-|n>^* = <n|\hat{R}^+|n-1> \tag{3.31}$$

which implies the relationship

$$(c_n^-)^* = c_{n-1}^+ \tag{3.32}$$

Next, notice that the ladder operators,

$$\hat{R}^\pm = \frac{1}{\sqrt{2}}\left[-i\hbar\frac{d}{dx} \pm i\omega x\right] \tag{3.33}$$

transform real functions into imaginary functions and vice versa. Since the harmonic oscillator eigenfunctions are real, the constants $c_n^\pm$ must be pure imaginary numbers. Finally, we have already seen in *Quantum mechanics 1*,

Section 3.2 (eqn 3.12) that $\hat{R}^+\hat{R}^- = \hat{H}^{(0)} - \frac{1}{2}\hbar\omega$, hence we have the matrix element

$$< n|\hat{R}^+\hat{R}^-|n > = n\,\hbar\omega \tag{3.34}$$

But this matrix element is also equal to $c_n^- c_{n-1}^+$, and according to the discussion above these constants are complex conjugates and pure imaginary. We therefore assign

$$c_n^- = -i\sqrt{n\,\hbar\omega} \tag{3.35}$$

$$c_n^+ = i\sqrt{(n+1)\,\hbar\omega} \tag{3.36}$$

which enables us to write for the matrix elements of $\hat{x}$

$$< n+1|\hat{x}|n > = \frac{1}{\omega\sqrt{2\mu}}\sqrt{(n+1)\,\hbar\omega} \tag{3.37}$$

$$< n-1|\hat{x}|n > = \frac{1}{\omega\sqrt{2\mu}}\sqrt{n\,\hbar\omega} \tag{3.38}$$

All other matrix elements are zero.

These matrix elements are now applied in eqn 3.24 for the calculation of the second order correction to the energy, which becomes

$$E_n^{(2)} = \frac{(VQ)^2 < n|\hat{x}|n-1 >< n-1|\hat{x}|n >}{E_n^{(0)} - E_{n-1}^{(0)}}$$
$$+ \frac{(VQ)^2 < n|\hat{x}|n+1 >< n+1|\hat{x}|n >}{E_n^{(0)} - E_{n+1}^{(0)}} \tag{3.39}$$

Using the matrix elements we have just calculated and the known energy levels of the harmonic oscillator, $E_n^{(0)} = (n+\frac{1}{2})\hbar\omega$ (see *Quantum mechanics 1*, eqn 3.7), we obtain the final result

$$E_n^{(2)} = (VQ)^2\left(\frac{n}{2\mu\omega^2} - \frac{n+1}{2\mu\omega^2}\right)$$
$$= -\frac{(VQ)^2}{2\mu\omega^2}$$
$$= -\frac{(VQ)^2}{2k} \tag{3.40}$$

Once the matrix elements are known the calculation is simple.

Exact solution. In this case the second-order corrected energy can be shown to be the exact solution to the problem. This can be seen by considering the unperturbed and the perturbed potential energy functions shown in Fig. 3.1. Both can be seen to be parabolas with the same curvature, but the minimum of the perturbed potential is translated to a coordinate $-VQ/k$ and the value of the potential energy at the minimum is $-(VQ)^2/2k$. The eigenfunctions of the perturbed potential are therefore given by the zero-order eigenfunctions expressed as functions of the variable $x + VQ/k$ and the corresponding eigenvalues are the zero-order eigenvalues shifted downwards by the quantity $(VQ)^2/2k$.

These results can be understood simply in terms of Fig. 3.1. Alternatively they can be demonstrated analytically. We simply outline the demonstration here. The perturbed equation is

$$-\frac{\hbar^2}{2\mu}\frac{\mathrm{d}^2\psi}{\mathrm{d}x^2} + \frac{1}{2}kx^2\psi + VQx\psi = E\psi \tag{3.41}$$

The potential energy is a quadratic function of x, so by completing the square we can transform the perturbed equation back into an unperturbed harmonic oscillator equation in terms of a transformed variable.

$$-\frac{\hbar^2}{2\mu}\frac{\mathrm{d}^2\psi}{\mathrm{d}x^2} + \frac{1}{2}k\left(x + \frac{VQ}{k}\right)^2\psi - \frac{(VQ)^2}{2k}\psi = E\psi \tag{3.42}$$

We now change the variable to $y = x + VQ/k$,

$$-\frac{\hbar^2}{2\mu}\frac{\mathrm{d}^2\psi}{\mathrm{d}y^2} + \frac{1}{2}ky^2\psi = \left(E + \frac{(VQ)^2}{2k}\right)\psi \tag{3.43}$$

The boundary conditions demand that ψ should tend to zero as x approaches $\pm\infty$, and since y is simply x plus a constant, the same conditions apply to y. The eigenfunctions are therefore simply the eigenfunctions of the harmonic oscillator, as functions of y, a coordinate with a shifted origin. We therefore conclude that the eigenvalues of this equation are the eigenvalues of the simple harmonic oscillator, that is

$$\left(E + \frac{(VQ)^2}{2k}\right) = (n + \tfrac{1}{2})\hbar\omega \tag{3.44}$$

so that

$$E_n = (n + \tfrac{1}{2})\hbar\omega - \frac{(VQ)^2}{2k} \tag{3.45}$$

which is identical to the result from second-order perturbation theory. The zero order and perturbed energy levels and eigenfunctions are depicted in Fig. 3.2.

Physically, the first-order change in the energy when an external electric field is applied is proportional to the dipole moment, and the second-order change is equal to $\frac{1}{2}\alpha V^2$, where α is the polarizability. It is clear from the results of this section that an oscillating charge in a harmonic potential has zero dipole moment, and its polarizability is equal to Q^2/k.

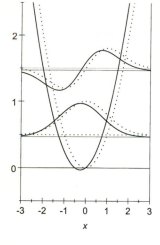

Fig. 3.2 The eigenfunctions and eigenvalues of the unperturbed potential (dotted lines), and the perturbed potential (full lines).

3.3   Rotation and vibration of a diatomic molecule

The full rotation–vibrational Hamiltonian for a diatomic molecule in the Born–Oppenheimer approximation is given in *Quantum mechanics 1*, eqn 2.39 as

$$\hat{H} = -\frac{\hbar^2}{2\mu}\left(\frac{\mathrm{d}^2}{\mathrm{d}r^2} + \frac{2}{r}\frac{\mathrm{d}}{\mathrm{d}r}\right) + V(r) + \frac{\hbar^2 J(J+1)}{2\mu r^2} \tag{3.46}$$

where r represents the bond length of the molecule and μ the reduced mass. The first two terms represent the kinetic energy and the potential energy for the motion of the bond length, that is, the vibrational motion. The final term is the rotational kinetic energy. The problem is simplified by a simple transformation, substituting $S(r) = r\psi(r)$, which yields

$$-\frac{\hbar^2}{2\mu}\frac{d^2 S}{dr^2} + V(r)S + \frac{\hbar^2 J(J+1)}{2\mu r^2}S = E_{vr}S \tag{3.47}$$

The zero-order approximation is the *rigid rotor harmonic oscillator* (RRHO) approximation, in which the potential energy $V(r)$ is assumed to be $\frac{1}{2}k(r - r_e)^2$ and the bond length variable in the rotational energy operator is replaced by the equilibrium bond length, r_e. This leads to a separation of the vibrational and rotational motions, and the rovibrational energy is given by

$$E_{vr}^{(0)} = \left(n + \frac{1}{2}\right)\hbar\omega + \frac{\hbar^2 J(J+1)}{2\mu r_e^2}$$

We can improve the RRHO approximation systematically by recognizing that if the degree of vibrational excitation is low, the vibrational wavefunction will be concentrated in the neighbourhood of the minimum of the potential energy curve $V(r)$. This suggests that we can perform a power expansion of the potential energy operator and the rotational energy operator about the equilibrium bond length, that is, use the variable $x = r - r_e$:

$$V(r) = V(r_e) + V'(r_e)x + \frac{1}{2}V''(r_e)x^2 + \frac{1}{6}V'''(r_e)x^3 + \dots$$
$$= \tfrac{1}{2}kx^2 - ax^3 + bx^4 + \dots \tag{3.48}$$

where k, a, and b are simply numbers have obvious interpretations in terms of the derivatives of V at the minimum. The first term in the expansion is zero because the minimum of the potential energy curve is taken to be the zero of the energy scale. The second term is zero because the minimum is a turning point and therefore the first derivative $V'(r_e)$ is zero.

$$\frac{\hbar^2 J(J+1)}{2\mu r^2} = \frac{\hbar^2 J(J+1)}{2\mu}\left(\frac{1}{r_e^2} - \frac{2x}{r_e^3} + \frac{3x^2}{r_e^4} - \frac{4x^3}{r_e^5} + \dots\right) \tag{3.49}$$

If we only take the first term in each power expansion, $\frac{1}{2}kx^2$ and $\hbar^2 J(J+1)/2\mu r_e^2$, respectively we obtain the RRHO approximation. Subsequent terms in the power expansion can then be used to correct the RRHO energy levels systematically using perturbation theory.

The harmonic oscillator approximation is successful because the leading term in a power expansion of the potential energy about its minimum is quadratic, thus the quadratic approximation is good in the neighbourhood of the minimum, that is, at low values of the vibrational quantum number, as illustrated in Fig. 3.3.

The RRHO approximation can be refined by extending the expansions to higher powers and using perturbation theory. Taking the first two correction

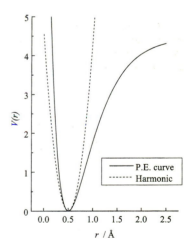

Fig. 3.3 The potential energy curve for H_2, and the first term in the power expansion for it (the harmonic approximation).

terms in each power expansion about r_e we obtain the perturbation Hamiltonian

$$-ax^3 + bx^4 - \frac{\hbar^2 J(J+1)x}{\mu r_e^3} + \frac{3\,\hbar^2 J(J+1)x^2}{2\mu r_e^4} \tag{3.50}$$

Physically, these corrections arise because the vibrational potential energy curve is not harmonic and because the bond stretches as the degree of vibration or rotation increases, thus increasing the moment of inertia and decreasing the rotational constant, which is inversely proportional to the moment of inertia.

The first correction terms in each expansion (the terms in x and x^3) are considered to be first-order corrections and the second correction terms (in x^2 and x^4) are second-order corrections. The use of perturbation theory will only be appropriate if these corrections are small; successive terms in the power expansion then give successively higher orders of approximation to the effective potential.

The matrix elements of powers of x required for the application of perturbation theory can be found by repeated application of the ladder operators used in the previous section. The results of such an analysis can be found in Table 3.1. Note that $< j|\hat{x}^k|n > = < n|\hat{x}^k|j >$ because the operator $\hat{x}^k$ is Hermitian and all the zero-order wavefunctions are real.

The first-order correction to the energy of level n requires the matrix elements $< n|\hat{x}|n >$ and $< n|\hat{x}^3|n >$, which can both be seen to be zero by inspection. x and x^3 are both odd functions of x, and $\psi_n^*\psi_n$ must be even, so that the integrand is odd and the integral vanishes identically.

The second-order correction is made up of two parts: the expectation of the second order perturbation and the second-order term from the first-order perturbation (see eqns 3.13 and 3.24). We deal with each separately.

The expectation of the second-order perturbation can be found from Table 3.1:

$$< n|\frac{3\,\hbar^2 J(J+1)x^2}{2\mu r_e^4} + b\hat{x}^4|n >$$

$$= \frac{3\,\hbar^2 \gamma^2(2n+1)J(J+1)}{2\mu r_e^4} + b\gamma^4 3(2n^2 + 2n + 1)$$

$$= \frac{3\,\hbar^3(n+\tfrac{1}{2})J(J+1)}{2\mu^2 r_e^4 \omega_e} + \frac{3b\hbar^2}{2\mu^2 \omega_e^2}[(n+\tfrac{1}{2})^2 + \tfrac{1}{4}] \tag{3.51}$$

The other contribution comes from the linear and cubic corrections to the Hamiltonian. Using eqn 3.24 we obtain

$$\sum_j{}' \left\{ \left[|< j| - \frac{\hbar^2 J(J+1)x}{\mu r_e^3} - ax^3|n >| \right]^2 / \hbar(n-j)\omega_e \right\} \tag{3.52}$$

Table 3.1 Harmonic oscillator matrix elements

$< n-1\|\hat{x}\|n >$	$=$	$\gamma\sqrt{n}$
$< n-2\|\hat{x}^2\|n >$	$=$	$\gamma^2\sqrt{n(n-1)}$
$< n\|\hat{x}^2\|n >$	$=$	$\gamma^2(2n+1)$
$< n-3\|\hat{x}^3\|n >$	$=$	$\gamma^3\sqrt{n(n-1)(n-2)}$
$< n-1\|\hat{x}^3\|n >$	$=$	$\gamma^3 3n\sqrt{n}$
$< n-4\|\hat{x}^4\|n >$	$=$	$\gamma^4\sqrt{n(n-1)(n-2)(n-3)}$
$< n-2\|\hat{x}^4\|n >$	$=$	$\gamma^4(4n-2)\sqrt{n(n-1)}$
$< n\|\hat{x}^4\|n >$	$=$	$\gamma^4 3(2n^2+2n+1)$

$\gamma = (\hbar/2\mu\omega)^{1/2}$. Matrix elements not related to these are zero.

The only non-zero matrix elements in the sum are those in which $j = n \pm 1$ and those in which $j = n \pm 3$, giving

$$\frac{|<n-3|ax^3|n>|^2}{3\hbar\omega_e} + \frac{|<n-1|\frac{\hbar^2 J(J+1)x}{\mu r_e^3}+ax^3|n>|^2}{\hbar\omega_e} +$$

$$\frac{|<n+1|\frac{\hbar^2 J(J+1)x}{\mu r_e^3}+ax^3|n>|^2}{-\hbar\omega_e} + \frac{|<n+3|ax^3|n>|^2}{-3\hbar\omega_e} \quad (3.53)$$

Substituting from Table 3.1 this becomes

$$E_{n,J}^{(2)} = -\frac{15a^2\hbar^2}{4\mu^3\omega_e^4}(n+\tfrac{1}{2})^2 - \frac{7a^2\hbar^2}{16\mu^3\omega_e^4}$$

$$-\frac{3a\hbar^3}{\mu^3 r_e^3\omega_e^3}(n+\tfrac{1}{2})J(J+1) - \frac{\hbar^4}{2\mu^3 r_e^6\omega_e^2}J^2(J+1)^2 \quad (3.54)$$

The correction to the energy can be combined with the zero-order energy and rewritten in the following simplified way, which helps to identify the origin of the various terms in the correction:

$$E_{n,J} = \hbar\omega_e(n+\tfrac{1}{2}) - \hbar\omega_e x_e(n+\tfrac{1}{2})^2$$

$$+ (B_e - \alpha_e(n+\tfrac{1}{2}))J(J+1) - D_e J^2(J+1)^2 \quad (3.55)$$

This is the empirical expression usually employed to analyse the rotational and vibrational structure of diatomic molecular spectra (Herzberg 1950). However, the perturbation analysis enables us to interpret each of the empirical constants in the energy level expression in terms of parameters describing the potential energy curve of the molecule.

The zero-order vibrational energy is that of the harmonic oscillator, $\hbar\omega_e(n+\tfrac{1}{2})$, and the zero-order rotational energy is $B_e J(J+1)$; the rotational constant $B_e = \hbar^2/2\mu r_e^2$. The constant x_e is called the *anharmonicity constant*, and is related to the corrections in the Hamiltonian by

$$\hbar\omega_e x_e = \frac{3\hbar^2}{2\mu^2\omega_e^2}\left[\frac{5a^2}{2\mu\omega_e^2} - b\right] \quad (3.56)$$

This constant contains only a and b from the perturbation, and can be thought of as correcting the vibrational energy for the anharmonicity in the potential. It accounts for the observation that adjacent vibrational energy levels are not equally spaced (cf. the harmonic approximation).

The constant D_e is the *centrifugal distortion constant*, given by:

$$D_e = \frac{4B_e^3}{\hbar^2 \omega_e^2} \qquad (3.57)$$

This term represents the effect of centrifugal force on the bond length. As the molecular rotation increases the bond stretches, as a result of the increased centripetal force required to hold the atoms together, thus reducing the apparent rotational constant. The effect is normally small, and is only important in accurate work and in 'pure rotational' spectroscopy. In infrared (vibration–rotation) spectroscopy the vibration–rotation coupling of the remaining term is usually far more important.

The constant α_e is known as the vibration–rotation coupling constant, and is given by

$$\alpha_e = 3B_e^2 \left[\frac{12a\hbar}{\mu\omega_e^2 \sqrt{2\mu B_e}} - 1 \right] \qquad (3.58)$$

This is a cross-term, showing that the effective rotational constant depends on the vibrational energy level: $B_n = B_e - \alpha_e(n + \frac{1}{2})$. In other words the apparent moment of inertia of a molecule depends on its vibrational state. The physical interpretation of this observation is that as vibrational excitation is increased, the amplitude of the vibration increases, and the moment of inertia increases, so decreasing the rotational constant, which is inversely proportional to the moment of inertia.

The spectroscopic effect of vibration–rotation coupling on the fine structure of a vibration–rotation band is important. Neglecting the centrifugal distortion,

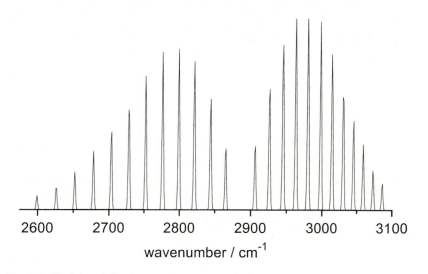

Fig. 3.4 The infrared vibration–rotation spectrum of HCl, illustrating the unequal spacing of the fine structure lines.

which is usually much smaller, we can use the energy level formula to find the wavenumber of lines in the infrared spectrum of a diatomic molecule. The energy difference for the $R(J)$ line ($J \leftarrow J + 1$) is

$$E_{1,J+1} - E_{0,J} = \hbar\omega_{\mathrm{e}}(1 - 2x_{\mathrm{e}}) + (2B_{\mathrm{e}} - \alpha_{\mathrm{e}}(J + 3))(J + 1) \qquad (3.59)$$

In the absence of vibration–rotation coupling, lines in the R branch would be equally spaced with a spacing $2B_{\mathrm{e}}$. However, when vibration–rotation coupling is included, the spacing diminishes as the J quantum number is increased. (It is easily calculated to be $2B_{\mathrm{e}} - \alpha_{\mathrm{e}}(2J + 5)$.) Similarly, the energy difference for the $P(J)$ line ($J \leftarrow J - 1$) is

$$E_{1,J-1} - E_{0,J} = \hbar\omega_{\mathrm{e}}(1 - 2x_{\mathrm{e}}) - (2B_{\mathrm{e}} + \alpha_{\mathrm{e}}(J - 2))J \qquad (3.60)$$

In the absence of vibration–rotation coupling, the lines in the P branch would also have an equal spacing $2B_{\mathrm{e}}$. However, when vibration–rotation coupling is included, the spacing increases as the J quantum number is increased. (It is easily calculated to be $2B_{\mathrm{e}} + \alpha_{\mathrm{e}}(2J + 1)$.) This unequal spacing is illustrated in Fig. 3.4. In spectroscopic terminology the band *degrades to the red*.

In writing these equations we have neglected a small constant term. Its inclusion simply shifts the origin of the energy scale slightly, affecting all energy levels to the same extent. It is physically unobservable since its magnitude is smaller than the precision to which electronic energies can be calculated. We shall therefore say no more about it.

3.4   Polyatomic molecules

The same methods can be applied to polyatomic molecules, except that the situation is more complicated because many degrees of freedom can now be coupled. The zero-order (RRHO) description has been given in Chapters 2 and 3 of *Quantum mechanics 1*: the rotations are described assuming that the molecule is rigid (*Quantum mechanics 1*, Section 3.4), and the harmonic vibrations are separated by a normal coordinate analysis (*Quantum mechanics 1*, Section 2.9).

The normal vibrational coordinates Q_j are such that the leading term of the Taylor expansion of the potential energy can be expressed

$$V = \sum_j Q_j^2 + \ldots \qquad (3.61)$$

There are no quadratic cross-terms in the Hamiltonian because the normal coordinates are calculated by diagonalizing the quadratic potential energy matrix. However, this expression is only valid if all the vibrational motions are harmonic, and furthermore we have assumed that the molecule is a rigid rotor. In reality neither of these will be true. We must therefore expect that the Hamiltonian should be corrected by terms such as

$$\sum_{ijk} a_{ijk}\hat{Q}_i\hat{Q}_j\hat{Q}_k \qquad (3.62)$$

$$\sum_{ijkl} b_{ijkl}\hat{Q}_i\hat{Q}_j\hat{Q}_k\hat{Q}_l \qquad (3.63)$$

which represent the anharmonic corrections to the vibrational potential energy.

There will also be centrifugal terms, involving combinations of vibrational coordinates and the angular momenta about the principal axes,

$$\sum_{irs} c_{irs} \hat{Q}_i \hat{J}_r \hat{J}_s \tag{3.64}$$

$$\sum_{ijrs} d_{ijrs} \hat{Q}_i \hat{Q}_j \hat{J}_r \hat{J}_s \tag{3.65}$$

and Coriolis terms, coupling rotational motion with vibrational momenta

$$\sum_{ijr} e_{ijr} \hat{Q}_i \hat{P}_j \hat{J}_r \tag{3.66}$$

The Coriolis force is the sideways force perceived when attempting to walk outwards from the centre of a rotating merry-go-round (carousel). It arises because the angular momentum of the body is constant; as you walk outwards your tangential velocity increases and you experience a sideways force in the rotating frame of reference: the foot that steps outwards moves sideways faster than the foot left behind on the inside. Nuclei in a molecule attempting to follow their normal vibrational motion while the molecule is rotating experience the same kind of disorienting effect.

There is evidently a large variety of possible terms to consider in any particular case, but many of the possible combinations of coordinates are absent from the Hamiltonian for symmetry reasons. As we have previously noted, within the Born–Oppenheimer approximation the Hamiltonian is totally symmetric with respect to all operations of the molecular point group. Since the Hamiltonian consists of a sum of terms, each of these must itself be totally symmetric. Neither the kinetic energy operator nor the potential energy operator can be altered by any symmetry operation of the molecule. The only possible combinations of normal modes, angular momentum operators, and so on, that can appear as perturbations in the Hamiltonian are therefore those that generate the totally symmetric irrep of the molecular point group.

For example in the CO_2 molecule, if we denote the normal coordinates as follows (see Table 3.2) then the only cubic anharmonicity terms that can appear in the Hamiltonian are the terms of Σ_g^+ symmetry

$$a_{111}\hat{Q}_1^3, \ a_{122}\hat{Q}_1\hat{Q}_2^2, \ a_{133}\hat{Q}_1(\hat{Q}_{3x}^2 + \hat{Q}_{3y}^2) \tag{3.67}$$

Table 3.2   Vibrations of CO_2

Q_1	symmetric stretch	Σ_g^+
Q_2	asymmetric stretch	Σ_u^+
Q_{3x}, Q_{3y}	bend	Π_u

References

Cohen-Tannoudji, C., Diu, B. and Laloë, F., (1992). *Mécanique quantique II.* Hermann, Paris.

Feynman, R. P. and Hibbs, A. R. (1965). *Quantum mechanics and path integrals.* McGraw-Hill, New York.

Green, N. J. B. (1997). *Quantum mechanics 1.* Oxford Chemistry Primer, Oxford.

Herzberg, G.(1950). *Spectra of diatomic molecules.* Van Nostrand Reinhold, New York.

4 Fields and couplings

4.1 Degenerate perturbation theory

In the perturbation theory presented in Chapter 3 the energy levels concerned are assumed to be non-degenerate. For the examples we have considered so far this has generally been true. However, many important chemical systems have degenerate states (for example molecular rotational energy levels or atomic energy levels), and the perturbation may have the effect of altering or removing the degeneracy. Perturbations of this kind arise if the molecule is subject to an external influence with a reduced symmetry, splitting a degeneracy in the unperturbed molecule, or if we have attempted to describe a system with a higher symmetry than is appropriate (e.g. if the molecule is distorted by the Jahn–Teller effect).

In Chapter 3 the results for the second-order energy, or for the first-order correction to the wavefunction, are expressed as expansions whose coefficients depend inversely on the zero-order energy differences between the basis states. These coefficients are not defined when the energy difference is zero, so the perturbation method is doomed to failure, unless it can be arranged that all coefficients linking degenerate zero-order states be identically zero. A related difficulty arises if the unperturbed states are nearly degenerate and the perturbation is larger than the energy difference between them. In this case the perturbation series does not contain undefined terms, but the perturbation causes large changes in the wavefunctions and the series solution may converge only at very high orders, or not at all.

A zero-order function in a degenerate space requires at least two labels to identify it unambiguously. Here we use n to label the energy level and m to distinguish the wavefunction from the other degenerate functions, that is, a typical zero-order function with energy $E_n^{(0)}$ is represented by the ket $|nm>$. However, there is a fundamental ambiguity in the choice of zero-order wavefunctions in a degenerate level because any linear combination of these wavefunctions is also an eigenfunction of the Hamiltonian with the same energy (see Problem 4.1.1). For example, in the hydrogen atom the three $2p$ orbitals may be chosen for their m_l values, which can be 1, 0, or -1, or alternatively can be chosen to point along the x, y, and z axes. Both choices give sets of three orthogonal, degenerate wavefunctions that span the same three-dimensional space, and there is an infinite choice of orthogonal sets of linear combinations that span the same space. However, if a perturbation is applied that lifts the degeneracy, each of the three eigenfunctions of the Hamiltonian is uniquely defined. For example, if the H atom is introduced into a magnetic field, the correct eigenfunctions are the p orbitals with defined m_l values. In general, the perturbed wavefunction will be close to just one of the infinite possible number of linear combinations of basis

Wavefunctions which are degenerate (or nearly so) can be strongly mixed by a perturbation and this may have a large effect on observable quantities such as spectral intensities and selection rules. An example in the vibrational spectra of polyatomic molecules is Fermi resonance.

Problem 4.1.1. *Show that if $|nl>$ and $nm>$ are degenerate eigenfunctions of the Hamiltonian, then all linear combinations of them $c_l|nl> + c_m|nm>$ are also eigenfunctions of the Hamiltonian with the same energy.*

functions, and this should be chosen as the appropriate zero-order wavefunction.

An alternative way to think about the problem is to start with the perturbed wavefunctions, then take the limit where the perturbation tends to zero. Each perturbed wavefunction will approach a characteristic linear combination of the zero-order functions. These combinations are the only suitable choices for a perturbation theory calculation, because a small perturbation will then only induce small corrections in the wavefunction. If any other combinations are chosen, the zero-order wavefunctions will be strongly mixed by the perturbation, however small it may be. In other words, zero-order functions should be chosen which recognize the symmetry of the perturbation. If this is done, all the offending coefficients are identically zero, and the problems of divergence do not arise.

The first step in constructing a perturbation theory for degenerate states must therefore be to find the correct linear combinations. Consider the zero-order energy level n with degeneracy g_n and energy $E_n^{(0)}$, and suppose that the correct choice for a zero-order wavefunction is given by

$$\psi_{nm}^{(0)} = \sum_j c_{mj}|nj> \tag{4.1}$$

where the $|nj>$ are a complete orthogonal set of zero order functions that span the degenerate space. The first-order perturbation from eqn 3.6, now becomes

$$(\hat{H}^{(0)} - E_n^{(0)})\psi_{nm}^{(1)} + (\hat{H}^{(1)} - E_n^{(1)})\sum_j c_{mj}|nj> = 0 \tag{4.2}$$

If we premultiply by a particular bra $< nk|$ and integrate we find

$$\sum_j c_{mj} < nk|\hat{H}^{(1)} - E_n^{(1)}|nj> = 0 \tag{4.3}$$

This is a linear equation linking all g_n coefficients c_{mj}. If we perform this operation with each of the g_n functions $< nk|$, we will obtain a different equation linking the coefficients in each case. We therefore obtain a set of g_n simultaneous equations for the coefficients.

The perturbation matrix elements are integrals that can be evaluated to give energies, and can be combined to form a matrix $\mathbf{H}^{(1)}$, whose elements are $H_{kj}^{(1)} = < nk|\hat{H}^{(1)}|nj>$. The coefficients c_{mj} can be similarly combined to form a vector $\mathbf{c}$. The set of simultaneous equations can then be written in matrix form:

$$(\mathbf{H}^{(1)} - E_n^{(1)}\mathbf{I})\mathbf{c} = \mathbf{0} \tag{4.4}$$

where $\mathbf{I}$ is the identity matrix and $\mathbf{0}$ is the zero vector. These homogeneous simultaneous equations, or *secular equations*, are equivalent to the secular equations derived from the variational method in Chapter 2. The solution will therefore give the best possible combination of the zero-order functions to approximate the ground state wavefunction. As discussed in Chapter 2, a non-trivial solution can only be found for $\mathbf{c}$ if the secular determinant is zero,

The correct linear combinations of degenerate zero-order functions are obtained by solving a set of *secular equations*.

$$|\mathbf{H}^{(1)} - E_n^{(1)}\mathbf{I}| = 0 \qquad (4.5)$$

since otherwise the inverse matrix would exist, and the only possible solution would be $\mathbf{c} = \mathbf{0}$.

Evaluation of the secular determinant leads to a polynomial of degree g_n for the energy correction $E_n^{(1)}$, which in general will have g_n solutions. Thus the first-order energy corrections are the eigenvalues of the matrix $\mathbf{H}^{(1)}$. Each value of the first-order energy correction gives a different set of secular equations, hence a different set of coefficients $\mathbf{c}$, that is, a different linear combination of the zero-order basis functions. Thus each perturbed energy level generates a zero-order wavefunction, which can be used as the basis of the perturbation expansion.

The unsuspecting reader might be fooled into thinking that because the perturbation Hamiltonian has been diagonalized, the eigenvalues represent the exact energies of the perturbed system. This is indeed sometimes the case, but it is not generally so, because the perturbation can also mix zero-order wavefunctions with different energies. In other words, $(\hat{H}^{(1)} - E_n^{(1)})|nj>$ may have components within the same degenerate subspace as $|nj>$, such as $c_{mk}|nk>$, but it may also have components with different energies, which are necessarily orthogonal to all the zero-order eigenfunctions with energy $E_n^{(0)}$(see *Quantum mechanics 1*, eqn 1.26). Diagonalization of the Hamiltonian matrix within a subspace ignores all components of the wavefunction outside the subspace and does not give the true energies in general. However, because the method results in secular equations for the coefficients, the variational principle guarantees that it gives the best possible approximate wavefunction within the restricted space spanned by the basis.

Solution of the secular equations gives a symmetry-adapted zero-order basis, in which the perturbation Hamiltonian is diagonal. The perturbation is not capable of mixing these symmetry-adapted degenerate zero-order states, and all the unpleasant terms in the perturbation expansion, which would otherwise have been infinite, are now identically zero. Once we have found this basis the usual perturbation expansion can be applied, with the deletion of all terms involving the mixing of degenerate states, so that the first-order correction to the wavefunction is

$$\psi_{nm}^{(1)} = \sum_{ij}{}' \frac{<\psi_{nm}^{(0)}|\hat{H}^{(1)}|\psi_{ij}^{(0)}>}{E_n^{(1)} - E_i^{(1)}} \psi_{ij}^{(0)} \qquad (4.6)$$

and the second-order energy correction is

$$E_n^{(2)} = \sum_{ij}{}' \frac{|<\psi_{nm}^{(0)}|\hat{H}^{(1)}|\psi_{ij}^{(0)}>|^2}{E_n^{(1)} - E_i^{(1)}} \qquad (4.7)$$

In both cases the sum excludes all zero-order states with energy $E_n^{(0)}$.

Of course, if the degeneracy is not removed in the first order, the secular equation analysis may be performed with a second-order perturbation, if there is one.

Doubly degenerate states

The foregoing discussion was fairly abstract, so we illustrate how simple the method can be to apply with the example of a doubly degenerate level where the correct answer is obvious from the start.

Consider a spinless particle with charge $-e$ confined to move around a ring in the xy plane. Each energy level of this system above zero is doubly degenerate. One choice of basis functions for energy level m is the pair $N\exp(\pm im\phi)$ with $N = 1/\sqrt{2\pi}$. Another equally valid choice could be $N'\sin m\phi$ and $N'\cos m\phi$ where $N' = 1/\sqrt{\pi}$. Let us start with the latter choice, which we denote $|s>$ and $|c>$.

Suppose that the system is perturbed by applying a magnetic field B in the z direction, which gives a perturbation Hamiltonian

$$\hat{H}^{(1)} = B\mu_B\hat{l}_z = -iB\mu_B\hbar\frac{\partial}{\partial\phi} \tag{4.8}$$

where $\hat{l}_z$ is the operator representing the z component of the angular momentum, $\hat{l}_z = -i\hbar\partial/\partial\phi$, see *Quantum mechanics 1*, Problem 1.3.3. It is elementary to apply this perturbation to the two basis functions to obtain

$$\hat{H}^{(1)}|s> \ = -iB\mu_B\,\hbar m|c>$$
$$\hat{H}^{(1)}|c> \ = iB\mu_B\,\hbar m|s>$$

giving the matrix elements

$$<s|\hat{H}^{(1)}|s> \ = 0$$
$$<c|\hat{H}^{(1)}|s> \ = -iB\mu_B\,\hbar m$$
$$<s|\hat{H}^{(1)}|c> \ = iB\mu_B\,\hbar m$$
$$<c|\hat{H}^{(1)}|c> \ = 0$$

The secular equations are therefore

$$-E^{(1)}c_c - iB\mu_B\,\hbar mc_s = 0$$
$$iB\mu_B\,\hbar mc_c - E^{(1)}c_s = 0$$

and the secular determinant

$$\begin{vmatrix} -E^{(1)} & -iB\mu_B\,\hbar m \\ iB\mu_B\,\hbar m & -E^{(1)} \end{vmatrix} = 0$$

Expanding the determinant and solving the resulting quadratic equation we find

$$E^{(1)} = \pm B\mu_B\,\hbar m$$

Taking the positive root and substituting into the secular equations we find $c_s = ic_c$ so that the normalized solution is

$$\psi_+^{(0)} = \frac{1}{\sqrt{2}}(|c> +i|s>) = \frac{1}{\sqrt{2\pi}}\exp(im\phi)$$

Similarly for the negative root we find $c_s = -ic_c$ so that the solution is $\exp(-im\phi)$.

Although we discovered the correct zero-order solutions the hard way, we could have gone through the following thought process instead. The zero-order states we want are those for which the perturbation matrix is diagonal, preferably states that are eigenfunctions of the perturbation, which will be the exact solutions to the problem. Since the perturbation operator is proportional to the z component of the angular momentum, the correct basis states have fixed values of this component. These states are $\exp(\pm im\phi)$ (see *Quantum mechanics 1*, Problem 1.3.4). The energy corrections in the first order are then the eigenvalues of the perturbation operator for these states, that is, $\pm B\mu_B \hbar m$; in fact these are the exact energy corrections for this problem.

Both methods give the same answer, of course, but this example illustrates two points: first the basic method required if the symmetry of the perturbation is not obvious; and second the great simplification that accrues if the symmetry of the perturbation can be recognized in advance. In the case of lifting true degeneracies this can often be done by using group theory to generate the correct linear combinations (*Quantum mechanics 1*, Section 4.7), but in the case of near degeneracies and accidental degeneracies we will always need to use the secular equations to find the correct zero-order states. The remainder of this chapter is devoted to analysing particular examples of this method that are relevant in chemistry. In most cases we use the symmetry of the problem to calculate the splittings induced by the perturbation; however, we do consider some systems where it is necessary to solve the secular equations.

4.2 External fields

Up to now we have only considered the ways in which interactions internal to an atom or molecule can be treated using perturbation theory. In each case, although an approximate degeneracy is lifted, we used a more exact symmetry to work out the states and energy levels to first order. In this section we use the same ideas to investigate the way in which an atom or molecule interacts with an externally applied perturbation, such as a magnetic field or an electric field.

Only non-zero magnetic moments can interact with magnetic fields. Atoms contain up to three sources of angular momentum – orbital motion, electron spin, and nuclear spin. In addition to these, molecules also have rotational angular momentum. The result is that most atoms and molecules have some form of magnetic moment, which interacts with an external magnetic field, giving rise to a measurable phenomenon, usually a splitting of lines in the spectrum.

- The magnetic moment from the electronic angular momentum of an atom gives rise to the Zeeman effect, in which absorption or emission lines in the atomic spectrum are split into multiplets by the application of the field, an effect which is taken advantage of in *laser magnetic resonance* spectroscopy, where the field is used to tune absorption lines into resonance with a laser.
- In a free radical, which has an electronic spin angular momentum associated with the unpaired electron, the magnetic field interacts with the

spin, splitting the energy levels in two. Transitions between these energy levels give rise to the phenomenon of *electron spin resonance* spectroscopy. Further splittings may arise as a result of the hyperfine interaction between the electron spin and the spins of magnetic nuclei in the molecule.

- In a closed shell molecule in the condensed phase the only source of magnetic moment is from the nuclear spins. The application of a magnetic field splits the molecular energy levels, and transitions between the resulting levels give rise to *nuclear magnetic resonance* spectroscopy.

In this section we shall confine our attention to systems which contain only a single source of angular momentum. The case where two sources of angular momentum are present requires a more detailed analysis, in which the coupling of the angular momenta is combined with the interaction Hamiltonian for the external field.

Zeeman effect

The *Zeeman* effect is the interaction of an atom with an external magnetic field.

A charged particle moving with an angular momentum (orbital, spin, or both) acts as a magnet, and will interact with any magnetic field that may be present. Magnetic fields may be externally applied, or arise from the motion of other charged particles in the atom or molecule. The classical energy of interaction between a magnet, whose magnetic moment is μ and a magnetic field, $\mathbf{B}$, is given by the scalar product $-\mu.\mathbf{B}$. For example, if $\mathbf{B}$ is an external field in the z direction, then the interaction is represented in the Hamiltonian by

$$\hat{H}_z = -\mu_z B \tag{4.9}$$

where μ_z is the z component of the magnetic moment.

The magnetic moment resulting from a particular motion is proportional to the angular momentum and to the charge of the particle, and inversely proportional to its mass. A particle with charge q and mass m rotating with an angular momentum $\mathbf{J}$ behaves like a magnet with a magnetic moment

$$\mu = g\mu_B \mathbf{J}/\hbar \tag{4.10}$$

where $\mu_B = \hbar q/2m$ and is known as the *Bohr magneton* for an electron, or the *nuclear magneton* for a nucleus. The appearance of the mass in the denominator, and the fact that atomic and molecular angular momenta have the order of magnitude of $\hbar$, means that magnetic effects will be of the order of 2000 times larger for electron motions than for nuclear motions. The g factor is more complex in origin: it takes the value 1 for electronic orbital motion, 2 for electron spin, and other values for nuclear spin, for example 5.585 for the proton spin. The Hamiltonian for the interaction of a magnetic moment with an external field now becomes

$$\hat{H}_z = -g\mu_B B \hat{J}_z/\hbar \tag{4.11}$$

In the absence of the field, each energy level has a definite total angular momentum quantum number J and degeneracy $2J + 1$. Fortunately we do not have to set up the secular equations to work out the effect of the field on these degenerate states because, if the field is weak, the Zeeman Hamiltonian is diagonal in the basis $|J, M_J>$. In other words the $2J+1$ states can be represented by assigning each of them a different value of M_J, the projection of the angular momentum in the z direction. Because the perturbation is proportional to the $\hat{J}_z$ operator, each of these states will be an eigenfunction of the Zeeman Hamiltonian, with eigenvalues

$$E^{(1)}_{JM_J} = -g\mu_B B M_J \qquad (4.12)$$

We conclude, therefore, that a level with angular momentum quantum number J is split by a magnetic field into $2J+1$ constituents, each having a characteristic value of M_J, and that the splitting is proportional to the magnetic field strength B.

For example, in a free radical with only electron spin $S = \frac{1}{2}$ to worry about, the energy levels are split into two, with $M_S = \pm\frac{1}{2}$. The g value is very close to 2 in this case, since the magnetic moment arises almost solely from the electron spin. Similarly in a borate anion, $^{11}BO_3^{3-}$, where the only angular momentum is the nuclear spin of the boron nucleus $\frac{3}{2}$, the energy levels are split into four by the application of the field (see Fig. 4.1).

We shall return to the Zeeman effect in Section 4.5, when we have discussed angular momentum coupling in atoms in more detail.

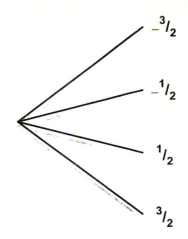

Fig. 4.1 The splitting of a nuclear spin state with $J = 3/2$ in an external magnetic field.

Stark effect

The Zeeman effect is found when an atom or molecule interacts with an external magnetic field. The *Stark effect* is a similar effect that arises when a species interacts with an external electric field. However, whereas the Zeeman effect is essentially a first-order effect, the Stark effect is usually zero in the first order, and provides an excellent example of the application of second-order perturbation theory.

The simplest case of the Stark effect is the effect of an electric field on the rotational energy levels of a heteronuclear diatomic molecule. The applied electric field interacts with the electric dipole moment of the molecule, and the strength of the interaction is proportional to the component of the dipole moment vector **d** in the direction of the field. If the electric field strength is denoted V and is taken to lie along the z direction, then the interaction Hamiltonian is

$$\hat{H}^{(1)} = Vd\cos\theta \qquad (4.13)$$

As discussed in *Quantum mechanics 1*, Chapter 3, the rotational wavefunctions of a diatomic molecule in the rigid rotor approximation are the spherical harmonics $Y_{JM}(\theta, \phi)$. The first-order modification in the energy is therefore

$$E^{(1)}_{JM} = Vd < JM|\cos\theta|JM > \qquad (4.14)$$

The properties of the spherical harmonics have been discussed in *Quantum mechanics 1*, Chapter 3, but they are well known, since they are the angular part of the atomic orbital functions. Each spherical harmonic has a well-defined behaviour with respect to the inversion symmetry operation. If the angular momentum quantum number J is even, the function is unchanged by inversion (cf. *s* and *d* orbitals), whereas if J is odd the function changes sign

The Stark effect is the interaction of an atom or molecule with an applied electric field.

on inversion (cf. p and f orbitals). If we now consider the integrand of the matrix element representing the first-order energy, we can see that the product $Y_{JM}^* Y_{JM}$ must always be even with respect to inversion, and since $\cos\theta$ is odd with respect to inversion, the integrand changes sign on inversion so that the integral must be zero.

The second-order energy is given by

$$E_{JM}^{(2)} = V^2 d^2 \sum_{JM}' \frac{|<JM|\cos\theta|J'M'>|^2}{E_{JM}^{(0)} - E_{J'M'}^{(0)}} \tag{4.15}$$

To apply this formula we need to know the zero-order rotational energy levels and the matrix elements of $\cos\theta$. The energy levels were shown in *Quantum mechanics 1*, Section 3.4 to be

$$E_{JM}^{(0)} = BJ(J+1) \tag{4.16}$$

where B is the rotational constant.

To calculate the matrix elements we use the fact that the angles θ and ϕ can be separated, as discussed in *Quantum mechanics 1*, Chapter 2.

$$Y_{JM}(\theta, \phi) = \Theta_{JM}(\theta)\Phi_M(\phi) \tag{4.17}$$

Because the perturbation does not contain the angle ϕ, the matrix element can be separated into the product of two integrals:

$$<JM|\cos\theta|J'M'> = <\Theta_{JM}|\cos\theta|\Theta_{J'M'}><\Phi_M|\Phi_{M'}> \tag{4.18}$$

Since Φ functions with different quantum numbers are orthogonal we can see immediately that the matrix element is zero if $M \neq M'$ so that we only need to find matrix elements with $M = M'$.

The integral over θ is a standard piece of applied mathematics which follows from the nature of the Θ functions (associated Legendre polynomials). The interested reader will find a discussion in Pauling and Wilson (1985). The result is that all matrix elements are zero apart from

$$<JM|\cos\theta|J-1, M> = \frac{J^2 - M^2}{(2J+1)(2J-1)} \tag{4.19}$$

and

$$<JM|\cos\theta|J+1, M> = \frac{(J+1)^2 - M^2}{(2J+3)(2J+1)} \tag{4.20}$$

Problem 4.2.1. *Given the zero-order rotational energy level formula (eqn 4.16) and the two non-vanishing matrix elements given in eqns 4.19 and 4.20, show that the second-order energy modification is*

$$E_{JM}^{(2)} = V^2 d^2 \frac{J(J+1) - 3M^2}{2BJ(J+1)(2J+3)(2J-1)} \tag{4.21}$$

We see that in the Stark effect the change in energy depends on the square of the quantum number M. The electric field does not therefore remove the degeneracy between wavefunctions with the same value of $|M|$ but with different signs. For example, Fig. 4.2 shows the effect of applying an electric field to the $J = 2$ rotational level.

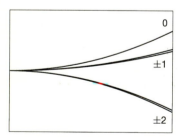

0

±1

±2

Fig. 4.2 The Stark effect on a rotational energy level with $J = 2$.

This result is extremely useful, since the Stark effect on the rotational spectrum can be used to give a very accurate measurement of the dipole moment of a molecule.

There are, of course, other energy levels of the molecule with the same values of J and M, but in different vibrational or electronic states, and strictly these should also be included in the perturbation sum. However, for most ground state diatomic molecules these levels are far away in energy from the level under consideration, and their contributions are very small. If deviations from the formula derived are observed, it is usually an indication that there is a nearby energy level in another electronic state with the correct values of J and M to give a significant contribution to the second-order energy. Spectroscopists use detailed analyses of such 'perturbations' to gain information on the relative energies of other electronic states.

A more complicated example of the second-order Stark effect is the ground state of the hydrogen atom. This has angular momentum quantum numbers $\ell = 0$ and $m = 0$ (instead of J and M). The only non-vanishing matrix elements in the second-order analysis are therefore those with $\ell = 1$ and $m = 0$, that is, the np_z orbitals. A full second-order analysis is given by Pauling and Wilson (1985).

The hydrogen atom in excited states displays a first-order Stark effect. Although we shall not calculate this effect in detail here, it is easy to see why it arises. In the hydrogen atom in state $n = 2$ there is an additional degeneracy. Not only are the three $2p$ orbitals degenerate, but the $2s$ orbital also has the same energy. The two $2p$ orbitals with $m = \pm 1$ obey the usual second-order Stark effect, but the matrix elements between the $2s$ and the $2p_z$ orbitals are non-zero, so we need to use degenerate perturbation theory. The secular determinant then takes the form

$$\begin{vmatrix} -E & \gamma V \\ \gamma V & -E \end{vmatrix} \tag{4.22}$$

where γ is a constant calculated from the matrix element mixing $2s$ and $2p_z$. The solutions for the energy levels are $\pm \gamma V$, giving an effect linear in the field strength ϵ, and the resulting wavefunctions are $\frac{1}{\sqrt{2}}(|2s> + |2p_z>)$ with energy γV and $\frac{1}{\sqrt{2}}(|2s> - |2p_z>)$ with energy $-\gamma V$. These sp hybrid wavefunctions are sketched in Fig. 4.3. In each of these zero-order states the hydrogen atom has a permanent dipole moment, of magnitude γ.

In the hydrogen atom the $2s$ orbital is stable, since the transition from $2s$ to $1s$ is forbidden by the parity selection rule, whereas transitions are permitted from $2p$ to $1s$. However, in the presence of an electric field the stationary states are mixtures of $2s$ and $2p_z$ so that the $2s$ wavefunction is 'contaminated' by $2p_z$ and the lifetime of an atom prepared in a $2s$ state is significantly shortened.

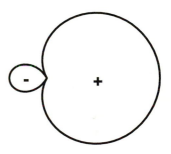

Fig. 4.3 An *sp* hybrid orbital for the linear Stark effect in the H atom.

4.3   Coupling of angular momenta

We shall deal with external fields in more detail in Section 4.5, but we must first understand the ways in which angular momenta interact magnetically with one another within an atom or a molecule, and the effects on the energy levels.

Many interactions in an atom or a molecule can be described as a coupling of angular momenta. They usually involve the vector addition of angular momenta, which are approximately conserved, to form a resultant angular momentum, which is more strongly conserved. For example, in Russell–Saunders coupling, the electronic orbital and spin angular momenta in an atom are separately coupled (summed) to make resultant vectors **L** and **S**, as discussed in Section 1.4. The vector addition of **L** with the total spin angular momentum **S** to form a resultant total electronic angular momentum **J** (spin–orbit coupling) is another example, discussed in detail shortly. Before we discuss the application of perturbation theory to assess the shifts in energy levels resulting from such couplings, which is the main subject of this section, we make a brief detour to discuss the rules for combining quantum mechanical angular momenta.

General rule for summing two angular momenta

In classical physics angular momentum is an axial vector quantity (see *Quantum mechanics 1*, p. 81). When angular momenta are summed, the problem simply amounts to their vector addition. As shown in Fig. 4.4, the two component vectors and their resultant form a triangle, thus the lengths of the vectors must obey the *triangle inequality*, that is, the sum of any two sides may not be smaller than the length of the remaining side, otherwise they could not join up to form a triangle. Thus if two vectors **L** and **S** are summed to give a resultant **J** then the following three inequalities for their lengths must be obeyed:

$$L + S \geq J$$
$$L + J \geq S$$
$$S + J \geq L$$

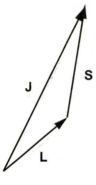

Fig. 4.4 The vector addition of **L** and **S** to give the total electronic angular momentum **J**.

The last two of these can be combined to give $J \geq |L - S|$. Thus the resultant vector can not be shorter than $|L - S|$, nor can it be longer than $L + S$. Although this analysis is entirely classical, the same restriction on the quantum numbers holds in quantum mechanics. If two angular momenta, whose quantum numbers are L and S are summed, the quantum number of the resultant cannot be smaller than $|L - S|$ or larger than $L + S$.

This result can be proved by considering the z components of the three angular momenta involved. Quantum mechanics only permits us to know one component of an angular momentum at a time (*Quantum mechanics 1*, Section 3.4). The z component of the resultant is simply obtained by summing the z components of the two constituent angular momenta,

$$M_J = M_L + M_S \tag{4.23}$$

We therefore consider all possible combinations of M_L and M_S and work out what value of M_J would result in each case. This gives us a set of possible values of M_J and we then simply have to work out which values of J could have given this set of M_J values.

The method is illustrated in Table 4.1 for the addition of $L = 2$ and $S = 1$. The M_L quantum number can take values from -2 to 2, the M_S quantum

Table 4.1 M_J values for summing angular momenta of magnitude 1 and 2

		M_S	
M_L	-1	0	1
2	1	2	3
1	0	1	2
0	-1	0	1
-1	-2	-1	0
-2	-3	-2	-1

number from -1 to 1. The table contains values of M_J for all possible combinations of these two quantum numbers. The largest value of M_J in the table is 3 (in general $L + S$), this must therefore be the largest value of J possible. A state with $J = 3$ will have 7 possible values of M_J, ranging from $+3$ to -3. If we remove one of each from the table, the largest remaining value of M_J is 2, there is therefore a state with $J = 2$, with M_J values from $+2$ to -2. Removing these from the table we are left with M_J values of $+1$, 0, and -1, that is, a state with $J = 1$. Thus the angular momenta with quantum numbers 1 and 2 add to give three possible resultants, with $J = 3$, $J = 2$, and $J = 1$.

Spin–orbit coupling

In Section 1.4 we considered the possible electronic energy levels of an atom on the assumption that the orbital and spin motion of the electrons can be separated from one another. This is the case in the non-relativistic Schrödinger equation, but in the relativistic theory of Dirac, where spin makes a natural appearance, several new factors appear, removing the degeneracy of the states in a term. The most important of these is *spin–orbit coupling*, whose physical origin can be understood by considering the orbital motion from the point of view of the electron. If the orbital motion has non-zero angular momentum, then in the frame of reference of the electron the nucleus moves around with a non-zero angular momentum, and therefore induces a magnetic field at the electron:

$$\mathbf{B} = -\frac{\mathbf{p}}{mc^2} \times \mathbf{E} \tag{4.24}$$

where $\mathbf{E}$ is the electric field through which the electron is travelling, which is the result of the Coulombic attraction of the nucleus and the spherically averaged repulsions of all the other electrons in the atom:

$$\mathbf{E} = -\frac{1}{er}\frac{dU}{dr} \mathbf{r} \tag{4.25}$$

U is the potential energy of the electron as a function of its distance from the nucleus. The relativistic origin of this term manifests itself in the appearance of the speed of light c. Combining these formulae,

$$\mathbf{B} = \frac{1}{emc^2 r}\frac{dU}{dr} \mathbf{p} \times \mathbf{r} = -\frac{1}{emc^2 r}\frac{dU}{dr} \mathbf{l} \tag{4.26}$$

where $\mathbf{p} \times \mathbf{r} = \mathbf{l}$ is the orbital angular momentum (*Quantum mechanics 1*, Section 3.4).

The electron has an intrinsic magnetic moment due to its spin, $\mu_s = es/m$ (note the g factor of 2), which interacts with this internal magnetic field,

$$-\boldsymbol{\mu}_s.\mathbf{B} = \frac{1}{m^2 c^2 r}\frac{dU}{dr}\mathbf{l} \cdot \mathbf{s} \tag{4.27}$$

Each electron in the atom will have such an interaction, leading to a perturbation Hamiltonian:

$$\hat{H}_{so} = \sum_i \alpha_i(r)\mathbf{l}_i.\mathbf{s}_i \tag{4.28}$$

where the summation is over all electrons.

Problem 4.3.1. *Use a similar construction to convince yourself that the addition of two angular momenta with quantum numbers L and S yields resultants whose quantum numbers lie between |L − S| and L + S in steps of one.*

Problem 4.3.2. *Show that the total degeneracy of a term, evaluated state by state, is equal to $(2L + 1)(2S + 1)$; i.e. that*
$$\sum_{J=|L-S|}^{L+S}(2J + 1) = (2L + 1)(2S + 1)$$
(Hint: this is the sum of an arithmetic progression.)

Spin–orbit coupling is the interaction of the spin magnetic moment of an electron with the magnetic field caused by its own orbital motion relative to the nucleus.

The presence of this term in the Hamiltonian means that the spin and orbital angular momenta can no longer be separated exactly, so the quantum numbers L and S are no longer exact. However, for light atoms the spin–orbit interaction is weak and can be treated as a small perturbation. Since the degeneracy of each term may be partially removed by the perturbation, we either need to set up secular equations, or else recognize the symmetry which leads to the degeneracies that are maintained. It is generally worth the effort to try and characterize the symmetry of the perturbation.

In the case of atomic states, although the two quantum numbers L and S are not conserved in the presence of spin–orbit coupling, the isotropy of space still requires that the total electronic angular momentum, J, be a conserved quantity. In a light atom the symmetry and electrostatic effects that lead to the coupling of the orbital angular momenta to give the resultant L and the spins to give the resultant S are much stronger than the spin–orbit coupling. Thus the total angular momentum $\mathbf{J}$ is given by the vector sum of the orbital and spin angular momenta:

$$\mathbf{J} = \mathbf{L} + \mathbf{S} \tag{4.29}$$

The possible values of the total angular momentum quantum number J are thus given by the vector sum rule, discussed in the previous subsection:

$$J = |L - S|, |L - S| + 1, \ldots, L + S \tag{4.30}$$

Since the total angular momentum J is a good quantum number we attempt to find an expression for the spin–orbit coupling Hamiltonian that will recognize this symmetry. The spin–orbit interaction operates primarily between the total spin angular momentum $\mathbf{S}$ and the total orbital angular momentum $\mathbf{L}$, and takes the form

$$\hat{H}_{so} = A\hat{\mathbf{L}}.\hat{\mathbf{S}} \tag{4.31}$$

The constant A is obtained by averaging the individual α_i over the states with the given values of L and S. It may be positive or negative, depending on whether the partly filled subshell is less or more than half full. To calculate the first-order energy correction we must find a basis in which the Hamiltonian is diagonal, that is, a way of expressing the Hamiltonian in terms of quantities whose eigenvalues are known. Fortunately this is easily done. Since $\mathbf{J}$ is the vector sum of $\mathbf{L}$ and $\mathbf{S}$ we have

$$\hat{J}^2 = (\hat{\mathbf{L}} + \hat{\mathbf{S}})^2 = \hat{L}^2 + \hat{S}^2 + 2\hat{\mathbf{L}}.\hat{\mathbf{S}}$$

so that

$$\hat{\mathbf{L}}.\hat{\mathbf{S}} = \frac{1}{2}(\hat{J}^2 - \hat{L}^2 - \hat{S}^2) \tag{4.32}$$

In the zero-order approximation all these operators have well-defined eigenvalues, so the first-order energy correction follows immediately

$$E_{LSJ}^{(1)} = \frac{A\hbar^2}{2}(J(J+1) - L(L+1) - S(S+1)) \tag{4.33}$$

Problem 4.3.3. *Show that the energy difference between states with quantum numbers J and J + 1 arising from the same term is A(J + 1). (The Lande interval rule.)*

Note that states arising from a given term have common values of L and S, so the splitting only depends on the quantum number J.

Although this analysis is sometimes quite accurate, the spin–orbit coupling Hamiltonian can also mix into a particular state contributions from other states with the same value of J but with different values of L and S, which can lead to important second-order energy corrections.

Problem 4.3.4. *What is the ground-state term symbol for the Si atom? This term is split into three energy levels by spin–orbit coupling, with energies 0 cm^{-1}, 77.14 cm^{-1}, and 223.31 cm^{-1} respectively. Identify the J values of these three states and comment on the energy spacings.*

Hyperfine coupling

In the preceding subsection we considered a term in the Hamiltonian which mixes the orbital motion of the electrons and their spins. There is a similar, but much smaller interaction between the electronic motion in an atom and the nuclear spin, known as the *hyperfine* interaction. Formally the hyperfine interaction comprises three separate interactions:

$$-\frac{\mu_0}{4\pi}\left(\frac{e}{mr^3}l.\boldsymbol{\mu}_s + \frac{3}{r}(\boldsymbol{\mu}_s.r)(\boldsymbol{\mu}_i.r) - \frac{1}{r^3}\boldsymbol{\mu}_s.\boldsymbol{\mu}_i + \frac{8\pi}{3}\boldsymbol{\mu}_s.\boldsymbol{\mu}_i\delta(\mathbf{r})\right) \quad (4.34)$$

where μ_0 is the permeability of space (a fundamental constant), the first term represents the interaction between the magnetic moment of the spinning nucleus $\boldsymbol{\mu}_s$ and the magnetic field due to the orbital motion of the electron. The second and third terms represent an interaction between the magnetic moments of the spinning electron and nucleus. The final term, known as the *Fermi contact interaction*, arises from the possibility that the electron may have non-zero density at the nucleus, where the magnetic field is very large. In practice, the final term is only important for s electrons, which have non-zero density at the nucleus, and the spherical symmetry of the electron distribution ensures that the other terms are identically zero. Since the only conserved electronic angular momentum in the atom is $\mathbf{J}$, the hyperfine interaction must be summed over all electrons in unfilled subshells and the electronic properties averaged for each value of J.

The procedure is rather involved, and finally yields (Landau and Lifshitz 1977, Cohen-Tannoudji *et al.* 1992) an interaction Hamiltonian,

$$\hat{H}_{hf} = A\hat{\mathbf{J}}.\hat{\mathbf{I}} \quad (4.35)$$

in which $\hat{\mathbf{J}}$ is the total electronic angular momentum operator and $\hat{\mathbf{I}}$ is the nuclear spin angular momentum operator. The coupling constant A can be either positive or negative.

The presence of a nuclear spin means that the total angular momentum is no longer $\mathbf{J}$; instead it will be a vector sum of the electronic and nuclear spin angular momenta, and in an isotropic space, this quantity (the total angular momentum) must be conserved:

$$\mathbf{F} = \mathbf{J} + \mathbf{I} \quad (4.36)$$

The hyperfine interaction is the interaction between the nuclear spin magnetic moment and the magnetic field induced by the electronic angular momentum.

The possible values of the total angular momentum quantum number F therefore range from a minimum of $|J - I|$ to a maximum of $J + I$. Each value of F will have a slightly different energy, which can be found in the same way as for spin–orbit coupling. Because the hyperfine interaction is very weak, J and I are still good quantum numbers and we can write

$$\mathbf{\hat{J}}.\mathbf{\hat{I}} = \frac{1}{2}(\hat{F}^2 - \hat{J}^2 - \hat{I}^2) \tag{4.37}$$

and therefore

$$E_{JIF}^{(1)} = \frac{A\hbar^2}{2}(F(F+1) - J(J+1) - I(I+1)) \tag{4.38}$$

Although this interaction is very small in the atomic spectrum, it is important for two reasons: first the observation of hyperfine structure in an atomic spectrum gives a useful method of measuring nuclear spin. Second, the hyperfine structure contains the major structural information about free radicals present in an electron spin resonance spectrum.

Angular momenta in diatomic molecules

[This section is more advanced.] In addition to electronic angular momenta and nuclear spin, diatomic molecules have rotational angular momentum. The nuclear spin effects are often neglected, and since most ground state molecules have closed shells, the resultant electronic angular momentum is zero. However, this is not the case for excited state molecules, and a few important molecules have unpaired electrons in their lowest configuration, notably NO and O_2.

One major difference between a molecule and an atom is that the potential energy for the electrons in a molecule is not spherically symmetrical. As a result, in a diatomic molecule, the orbital angular momentum L has a fixed projection on to the internuclear axis, usually labelled with the quantum number Λ. The electronic term symbol identifies the value of Λ: thus a Σ state has $\Lambda = 0$, Π signifies $\Lambda = 1$, Δ, $\Lambda = 2$, and so on.

The quantum mechanics of the angular momentum coupling and, therefore, the fine structure of the molecular energy levels depends on whether the spin–orbit coupling is greater or less than the differences between adjacent rotational energy levels. Molecules with large spin–orbit coupling are generally referred to as case (a), and molecules with small spin–orbit coupling as case (b). These two cases are limits and there are many examples of intermediate cases. Indeed a molecule may be well described by case (a) at low rotational energies, but undergo a transition to case (b) at higher energies, where rotational energy levels are more widely spaced. We only give a brief outline of the relevant angular momenta for these cases. The interested reader may pursue these matters in Herzberg (1950) or Landau and Lifshitz (1977). In both cases the zero-order description of the molecule has an orbital angular momentum Λ projected on to the internuclear axis, but neither spin nor rotational angular momentum are coupled in.

Hund's case (a).   In case (a) we start by neglecting rotation and couple the spin to the orbital angular momentum. The angular momenta involved are illustrated in Fig. 4.5. Because the orbital angular momentum is strongly

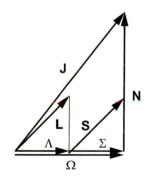

Fig. 4.5 The angular momenta involved in Hund's case (a).

coupled to the internuclear axis, the spin–orbit coupling $A(r)\mathbf{L.S}$ now becomes $\Lambda(r)\hbar^2\Lambda\Sigma$, where Σ is the quantum number representing the component of the spin parallel to the bond, and the bond length is assumed constant and equal to r. Σ can take values from $-S$ to $+S$ in steps of 1, in accord with the general rule for the projection of an angular momentum vector (*Quantum mechanics 1*, Section 3.4).

The component of the vector sum $\mathbf{L} + \mathbf{S}$ along the bond takes the value $\Lambda + \Sigma = \Omega$. The quantum number Ω can take values from $\Lambda + S$ to $\Lambda - S$, for example for the ground state of NO, $\Lambda = 1$ and $S = 1/2$, so that Ω can take the value $1/2$ or $3/2$. These electronic states are separated by about 200 cm^{-1} and are given the term symbols $^2\Pi_{1/2}$ and $^2\Pi_{3/2}$. Note that, unlike J in atoms, Ω can take negative values, for example a $^4\Pi$ term can have Ω values from $5/2$ to $-1/2$, and each will have a different energy. The shift in energy is therefore approximated to first order as $A(r)\Lambda\Sigma$.

We now have to consider the total angular momentum of the molecule, J, which is the vector sum of the orbital, the spin, and the rotational angular momenta. Since the rotational angular momentum $\mathbf{N}$ is perpendicular to the internuclear axis and the molecule has an electronic angular momentum along the bond Ω, the total angular momentum of the molecule must be at least Ω (one of its components is Ω). Hence J takes all values from Ω upwards in steps of 1.

The rotational angular momentum of the molecule is $\mathbf{J} - (\mathbf{L} + \mathbf{S})$ and so the operator for the rotational energy of the molecule is $B(r)(\hat{\mathbf{J}}^2 - (\hat{\mathbf{L}} + \hat{\mathbf{S}})^2$, where $B(r)$ is the usual rotational constant. Averaging over the zero-order wavefunctions (see Landau and Lifshitz 1977) we obtain

$$E_{\text{rot}} = B(r)\hbar^2[J(J+1) - 2\Omega^2 + \Lambda^2 + 2\Lambda\Sigma + S(S+1)] \qquad (4.39)$$

In a given term with a particular value of Ω, the rotational energy is proportional to $J(J + 1)$, which at first sight seems to be the same result as the normal closed shell molecule. However, it should be remembered that the values of J start at $|\Omega|$. If Ω is a non-zero integer, this will lead to 'missing' rotational lines. For example, the first line in the R branch of a vibrational transition within a $^1\Pi$ state is $R(1)$, and the first line in the P branch is $P(2)$. On the other hand, if Ω is a half integer, the J quantum numbers will also be half integers. For example, any attempt to analyse rotational fine structure in the spectrum of NO using integer values for the J quantum numbers is doomed to failure.

Hund's case (b). In case (b) the spin–orbit coupling is added last. The rotation of the molecule couples with the orbital angular momentum to give $\mathbf{K}$, whose quantum number can take values starting from Λ in steps of 1. The rotational energy is then $B(r)[K(K + 1) - \Lambda^2]$. The spin is then added to $\mathbf{K}$ to give a total molecular angular momentum $\mathbf{J}$. The J quantum number can take any value from $|K - S|$ to $K + S$, in steps of 1. The calculation of the spin–orbit splitting is illustrated in Fig. 4.6. The spin–orbit coupling operator is $A(r).\mathbf{S}$, but the component of $\mathbf{S}$ in the direction of Λ is not defined because Λ is strongly coupled to the rotation, forming $\mathbf{K}$, and $\mathbf{S}$ is then coupled to $\mathbf{K}$.

Problem 4.3.5. *Find the magnetic moment of the NO molecule in the $^2\Pi_{1/2}$ and $^2\Pi_{3/2}$ states. [Remember the g factor for the spin. In one state the spin and orbit contributions reinforce one another, whereas in the other state they act in opposite directions.]*

Problem 4.3.6. *The rotational constant of NO is 1.70 cm^{-1}. Predict the positions of the first two lines in the P and R branches of the vibrational spectrum relative to the band origin. How would these differ for a 'normal' closed-shell molecule with the same rotational constant?*

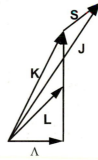

Fig. 4.6 The angular momenta involved in Hund's case (b).

Thus, the vector **S** does not have a well-defined direction, and precesses around **K**. Its component in the direction of **K** is $(\mathbf{S}.\mathbf{K})\mathbf{K}/K(K+1)$ and the spin–orbit coupling operator becomes

$$\hat{H}_{\mathrm{so}} = A(r)\frac{(\hat{\mathbf{\Lambda}}.\hat{\mathbf{K}})(\hat{\mathbf{S}}.\hat{\mathbf{K}})}{\hbar^2 K(K+1)} \tag{4.40}$$

Because $\mathbf{J} = \mathbf{K} + \mathbf{S}$, and $\mathbf{\Lambda}.\mathbf{K} = \hbar^2\Lambda^2$, this can be rewritten

$$\hat{H}_{\mathrm{so}} = A(r)\frac{\Lambda^2}{K(K+1)}\frac{1}{2}[\hat{\mathbf{J}}^2 - \hat{\mathbf{K}}^2 - \hat{\mathbf{S}}^2] \tag{4.41}$$

which has the eigenvalues

$$E_{\mathrm{so}} = A(r)\hbar^2\frac{\Lambda^2[J(J+1) - S(S+1) - K(K+1)]}{2K(K+1)} \tag{4.42}$$

For further details, the reader should consult Herzberg (1950) or Landau and Lifshitz (1977).

4.4   Nuclear magnetic resonance

In NMR spectroscopy we are concerned with transitions between nuclear spin states that have different characteristic energies in the presence of a large external magnetic field. The magnetic field is partially screened by orbital motion of the electrons induced by the field (similar to eddy currents) so that nuclei in electronically distinct environments experience slightly different local fields. The nuclear spin wavefunction of the molecule is characterized by specifying for each nucleus i a spin quantum number I_i and its z component M_i. In the absence of a magnetic field, all of these possible zero-order wavefunctions are degenerate. When the field is applied, if all of the nuclei experience different local magnetic fields, then each wavefunction has a different energy.

Electromagnetic radiation (usually radio-frequency) induces transitions between these states, with emission or absorption of photons. The only allowed transitions are those in which M for one nucleus changes by ± 1 and all other quantum numbers remain the same. If all the nuclei interact with the field differently, they absorb radiation with their own characteristic frequency. This frequency, expressed relative to a suitable standard, is called the *chemical shift*.

The situation is complicated by the interaction, or coupling, of nuclear spins, which shifts the energies of the wavefunctions further. The great utility of NMR, of course, is that the chemical shifts give information about the electronic environment of the nucleus, and the patterns induced by coupling give information about which nuclei are close to one another.

Since two perturbations are present, we can identify two limiting cases: either energy differences induced by the field are much greater than those induced by coupling, or *vice versa*. Both limiting cases are very important and need to be considered before looking at the more general situation where the effects are comparable in magnitude.

The AX spectrum

The first case we consider is where the dominant effect is the chemical shift difference, so that in the presence of the field the correct zero-order basis consists of the states characterized by the I and M quantum numbers for each nucleus. The Hamiltonian for the magnetic field effect is

$$\hat{H}_z = -\sum_i \delta_i B \hat{I}_{iz}/\hbar \tag{4.43}$$

where $\delta_i = (1 - \sigma_i)\mu_N g_i$, σ_i is the screening constant of the nucleus i and g_i is its g value (cf. eqn 4.10). The first-order energy level shift for a given state is then

$$E^{(1)} = -\sum_i B(1 - \sigma_i)\mu_N g_i M_i \tag{4.44}$$

The spin–spin coupling Hamiltonian has the same form as the angular momentum couplings discussed in Section 4.3, and is given by

$$\hat{H}_{ss} = \sum_{i>j} J_{ij} \hat{\mathbf{I}}_i . \hat{\mathbf{I}}_j/\hbar^2 \tag{4.45}$$

where J_{ij} is the coupling constant. Using the zero-order states appropriate for the field effect, the only component of the dot product which survives is the z component. (See later for a detailed discussion of this point.) The first-order energy level shifts are then

$$E^{(1)} = \sum_{i>j} J_{ij} M_i M_j \tag{4.46}$$

We illustrate this analysis by considering the case of a molecule with two magnetic nuclei, both of spin $I = \frac{1}{2}$, and with very different chemical shifts. This spectrum is called an AX-type spectrum. The four possible wavefunctions and the corresponding energy levels are

$$|\beta\beta> : \quad \tfrac{1}{2}(\delta_A + \delta_X)B \quad +\tfrac{1}{4}J_{AX}$$

$$|\beta\alpha> : \quad \tfrac{1}{2}(\delta_A - \delta_X)B \quad -\tfrac{1}{4}J_{AX}$$

$$|\alpha\beta> : \quad \tfrac{1}{2}(-\delta_A + \delta_X)B \quad -\tfrac{1}{4}J_{AX}$$

$$|\alpha\alpha> : \quad \tfrac{1}{2}(-\delta_A - \delta_X)B \quad +\tfrac{1}{4}J_{AX}$$

These energy levels, before and after application of the spin–spin coupling, are represented in Fig. 4.7. The four allowed transitions are also depicted, and their frequencies are tabulated in Table 4.2. Each transition has the same intensity, since the dipole moment matrix elements governing the allowed transitions (i.e. $< \alpha\alpha|\hat{\mu}|\alpha\beta >$, $< \alpha\alpha|\hat{\mu}|\beta\alpha >$, $< \alpha\beta|\hat{\mu}|\beta\beta >$, and $< \beta\alpha|\hat{\mu}|\beta\beta >$) are all equal. Both the other possible matrix elements ($< \alpha\alpha|\hat{\mu}|\beta\beta >$ and $< \beta\alpha|\hat{\mu}|\alpha\beta >$) are zero, because they involve simultaneous flipping of two spins. The resulting spectrum is depicted at the bottom of the figure. Each line with a characteristic chemical shift is split into two by

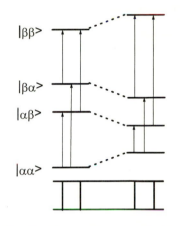

Fig. 4.7 The energy levels involved in an AX NMR spectrum. On the left are the levels without spin–spin coupling, on the right the levels have shifted up or down by $J/4$ to illustrate the effect of coupling. Underneath is a representation of the spectrum.

Problem 4.4.1. *The HCCD spectrum. Find expressions for the energy levels and the frequencies in the H NMR spectrum and the D NMR spectrum of the deuteroacetylene molecule. ($I_H = \frac{1}{2}$ and $I_D = 1$).*

Table 4.2   NMR AX spectrum frequencies

	$\alpha\alpha$	$\alpha\beta$	$\beta\alpha$	$\beta\beta$
$\alpha\alpha$		$B\delta_X + \frac{1}{2}J_{AX}$	$B\delta_A + \frac{1}{2}J_{AX}$	
$\alpha\beta$	$B\delta_X + \frac{1}{2}J_{AX}$			$B\delta_A - \frac{1}{2}J_{AX}$
$\beta\alpha$	$B\delta_A + \frac{1}{2}J_{AX}$			$B\delta_X - \frac{1}{2}J_{AX}$
$\beta\beta$		$B\delta_A - \frac{1}{2}J_{AX}$	$B\delta_X - \frac{1}{2}J_{AX}$	

A blank entry denotes that the transition is forbidden.

Problem 4.4.2. *The AMX spectrum. Repeat the analysis of this section for a molecule with three spin $\frac{1}{2}$ nuclei, each of which has a different chemical shift. If the three coupling constants are $J_{AM} = 10$ Hz, $J_{AX} = 6$ Hz, and $J_{MX} = 2$ Hz, find the energy levels and sketch the resulting spectrum. What differences would there be if the coupling constants were all negative?*

coupling with the other nucleus. The splitting in the fine structure of each line is equal to the coupling constant J_{AX}.

Equivalent nuclei

Many molecules contain several nuclei in identical environments, having identical chemical shifts and all coupling constants equal. In such a case the magnetic field fails to remove the degeneracy completely. For example in a system with two equivalent nuclei the two zero-order states $\alpha\beta$ and $\beta\alpha$ have exactly the same energy in the presence of the field. The effect of the spin–spin coupling then has to be included using degenerate perturbation theory. However, it is possible to solve this problem by symmetry, using methods similar to those of Section 4.3.

The Zeeman Hamiltonian is now

$$\hat{H}_z = -B\delta \sum_i \hat{I}_{iz} \tag{4.47}$$

Each of the component $\hat{I}_z$ operators has the same weight in the sum, suggesting that the resultant total nuclear spin angular momentum,

$$\mathbf{F} = \sum_i \mathbf{I}_i \tag{4.48}$$

and its z component F_z are conserved.

This suggestion is supported by the observation that in the absence of the field the total nuclear spin angular momentum is conserved regardless of any coupling between the spins. It is therefore possible to express the spin–spin coupling Hamiltonian in terms of the $\hat{F}^2$ and the $\hat{F}_z$ operators.

The coupling between equivalent nuclear spins has the Hamiltonian

$$\hat{H}_{ss} = J \sum_{i>j} \hat{\mathbf{I}}_i . \hat{\mathbf{I}}_j / \hbar^2 \tag{4.49}$$

because all the coupling constants are equal. This can be related to the operator for the total nuclear spin by considering the operator $\hat{F}^2$:

$$\hat{F}^2 = \sum_{i,j} \hat{\mathbf{I}}_i . \hat{\mathbf{I}}_j$$

$$= \sum_i \hat{I}_i^2 + 2 \sum_{i>j} \hat{\mathbf{I}}_i . \hat{\mathbf{I}}_j \tag{4.50}$$

Hence the spin–spin coupling Hamiltonian becomes

$$H_{ss} = \frac{J}{2\hbar^2}(\hat{F}^2 - \sum_i \hat{I}_i^2) \tag{4.51}$$

If the system contains n equivalent nuclei, each of which has spin I, any wavefunction with total nuclear spin angular momentum quantum number F and z component M will simultaneously be an eigenfunction of both the Zeeman perturbation and the spin–spin coupling Hamiltonian. The energy shift will be

$$E^{(1)} = B\delta M + \frac{J}{2}(F(F+1) - nI(I+1)) \tag{4.52}$$

As a simple example we consider the spectrum of a system of two identical spin $\frac{1}{2}$ nuclei, such as the formaldehyde molecule, H_2CO. Two spins of $\frac{1}{2}$ can couple to give total spin quantum numbers of 0 or 1, as discussed in Section 4.2. The wavefunctions and their quantum numbers are tabulated in Table 4.3.

Problem 4.4.3. *Use the equality of the matrix elements discussed in the last subsection to show that the only allowed transitions between $|F, M>$ states in the A_2 spectrum are the transitions $|1, 1> \leftrightarrow |1, 0>$ and $|1, 0> \leftrightarrow |1, -1>$. Further, show that each of these transitions has the same frequency and that the total intensity at this frequency is four times that of one of the lines in the AX spectrum. The energy levels and transitions are depicted in Fig. 4.8.*

The result derived in this problem is an example of a more general rule for equivalent nuclei, that the absorption or emission of electromagnetic radiation cannot alter the total nuclear spin angular momentum, but can alter its z component by only one unit.

This analysis enables us to explain a common misconception in the analysis of NMR spectra. It is often said that equivalent nuclei do not couple with one another. This claim is clearly not true, as the analysis leading to eqn 4.52 shows. However, the only allowed transitions take place between levels with the same value of F and adjacent values of M, and we can see from the equation that the effect of spin–spin coupling on the energy of each of these states is identical, so that the spacing is simply $B\delta$, and all the transitions occur with the same frequency. Although spin–spin coupling does affect the energy levels, it does not result in any splitting of the line in the NMR spectrum.

The AB spectrum

We now turn to the case where the energy difference between the $\alpha\beta$ and the $\beta\alpha$ basis states of two spin $\frac{1}{2}$ nuclei induced by the external field is of the same order of magnitude as the effects of spin–spin coupling. It is no longer possible to appeal to symmetry to find combinations of these basis states which are exact eigenfunctions of the Hamiltonian, the only way to find the correct spin wavefunctions and the corresponding energy levels is to apply both perturbations at once and diagonalize the matrix representing the effect of the Hamiltonian on the basis states, as discussed in Section 4.1.

Table 4.3 Wavefunctions for two equivalent spin $\frac{1}{2}$ nuclei.

F	M	ψ
1	1	$\alpha\alpha$
1	0	$\frac{1}{\sqrt{2}}(\alpha\beta + \beta\alpha)$
1	−1	$\beta\beta$
0	0	$\frac{1}{\sqrt{2}}(\alpha\beta - \beta\alpha)$

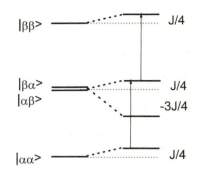

Fig. 4.8 The energy levels involved in an A_2 NMR spectrum. On the left are the levels without spin–spin coupling, on the right the levels have shifted to illustrate the effect of coupling.

Problem 4.4.4. *Calculate the energy levels and spectrum for an A_2X molecule, which contains two identical spin $\frac{1}{2}$ A nuclei and a spin $\frac{1}{2}$ X nucleus with a very different chemical shift.*

Problem 4.4.5. *Use the same method to calculate the NMR spectrum of an A_3X molecule, such as the proton spectrum of ethanal, CH_3CHO. (The coupling of three equivalent spin $\frac{1}{2}$ nuclei gives one state with $F = \frac{3}{2}$ and two states with $F = \frac{1}{2}$.)*

The first step is to construct the Hamiltonian matrix. To do this we consider the effect of the Hamiltonian on each of the four basis states, $|\alpha\alpha>$, $|\beta\alpha>$, $|\alpha\beta>$, and $|\beta\beta>$. The effects of the magnetic field are straightforward:

$$\hat{H}_z|\alpha\alpha> = -\frac{1}{2}B(\delta_A + \delta_B)|\alpha\alpha>$$

$$\hat{H}_z|\alpha\beta> = -\frac{1}{2}B(\delta_A - \delta_B)|\alpha\beta>$$

$$\hat{H}_z|\beta\alpha> = \frac{1}{2}B(\delta_A - \delta_B)|\beta\alpha>$$

$$\hat{H}_z|\beta\beta> = \frac{1}{2}B(\delta_A + \delta_B)|\beta\beta>$$

since the basis functions are eigenfunctions of the Zeeman Hamiltonian.

The effect of spin–spin coupling requires a little more care, however. The Hamiltonian is

$$\hat{H}_{ss} = J\hat{\mathbf{I}}_A.\hat{\mathbf{I}}_B = J(\hat{I}_{Ax}\hat{I}_{Bx} + \hat{I}_{Ay}\hat{I}_{By} + \hat{I}_{Az}\hat{I}_{Bz}) \qquad (4.53)$$

The z component of this operator is straightforward, but to work out the effects of the x and y components we need to use the ladder operators $\hat{I}_{\pm}$ introduced in *Quantum mechanics 1*, Section 3.4, and defined in eqn 3.26.

Problem 4.4.6.   *Given the definitions of $\hat{I}_+$ and $\hat{I}_-$, prove that*

$$\hat{\mathbf{I}}_A.\hat{\mathbf{I}}_B = \frac{1}{2}(\hat{I}_{A+}\hat{I}_{B-} + \hat{I}_{A-}\hat{I}_{B+}) + \hat{I}_{Az}\hat{I}_{Bz}$$

The effect of the ladder operators on a single spin can be summarized as follows:

$$\hat{I}_+|\alpha> = 0$$
$$\hat{I}_-|\alpha> = |\beta>$$
$$\hat{I}_+|\beta> = |\alpha>$$
$$\hat{I}_-|\beta> = 0$$

$\hat{I}_{\pm}$ converts a spin function with z component M to one with z component $M \pm 1$. Since there is no spin function with a spin $\frac{1}{2}$ and with z components larger than $\frac{1}{2}$, $\hat{I}_+|\alpha> = 0$ and $\hat{I}_-|\beta> = 0$. The ladder operators $\hat{I}_{A\pm}$ should be understood to operate only on the spin of nucleus A, and similarly for $\hat{I}_{B\pm}$ and nucleus B.

The operation of $\hat{H}_{ss}$ on the basis functions can therefore be summarized as follows:

$$\hat{H}_{ss}|\alpha\alpha> = \frac{1}{4}J|\alpha\alpha>$$

$$\hat{H}_{ss}|\alpha\beta> = \frac{1}{2}J|\beta\alpha> - \frac{1}{4}J|\alpha\beta>$$

$$\hat{H}_{ss}|\beta\alpha> = \frac{1}{2}J|\alpha\beta> - \frac{1}{4}J|\beta\alpha>$$

$$\hat{H}_{ss}|\beta\beta> = \frac{1}{4}J|\beta\beta>$$

The two basis states $|\alpha\alpha>$ and $\beta\beta>$ are clearly eigenfunctions of both the Zeeman and the spin–spin coupling Hamiltonians, but the states $|\alpha\beta>$ and $|\beta\alpha>$ are mixed by the spin–spin coupling perturbation.

The correct choices for the wavefunctions are superpositions of the form $c_1|\alpha\beta> +c_2|\beta\alpha>$, and the coefficients c_1 and c_2 must be found by solving the secular equations:

$$(\frac{1}{2}B\delta - \frac{1}{4}J - E)c_1 + \frac{1}{2}Jc_2 = 0 \qquad (4.54)$$

$$\frac{1}{2}Jc_1 - (\frac{1}{2}B\delta + \frac{1}{4}J + E)c_2+ = 0 \qquad (4.55)$$

where $\delta = \delta_B - \delta_A$.

These simultaneous equations only have non-zero solutions for c_1 and c_2 if the secular determinant is zero.

Substituting E_+ into the secular equations we obtain

$$\frac{c_2}{c_1} = \frac{\Delta - B\delta}{J}$$

Since the final wavefunction must be normalized we must also have

$$c_1^2 + c_2^2 = 1 \qquad (4.57)$$

We are therefore justified in expressing c_1 and c_2 in terms of a parameter θ such that

$$c_1 = \cos\theta$$
$$c_2 = \sin\theta$$

$\theta = \tan^{-1}[(\Delta - B\delta)/J]$, and if J is positive, θ lies between 0 and $\pi/4$. Note that θ depends only on the ratio between δB and J; the effect of increasing the magnetic field is therefore to reduce the value of θ.

Substituting E_- into the secular equation we find $c_1/c_2 = -\tan\theta$, so that

$$\psi_- = -\sin\theta|\alpha\beta> +\cos\theta|\beta\alpha> \qquad (4.58)$$

The transition frequencies are tabulated in Table 4.4. The line intensities are easily found from the constituent matrix elements discussed in connection with the AX spectrum.

Problem 4.4.7. *Prove that the secular determinant is only zero for the two values of E*

$$E_\pm = -\frac{1}{4}J \pm \frac{1}{2}\Delta \qquad (4.56)$$

where $\Delta = \sqrt{B^2\delta^2 + J^2}$

Table 4.4   NMR AB spectrum frequencies measured from the mean chemical shift of the pair.

	$\alpha\alpha$	ψ_+	ψ_-	$\beta\beta$
$\alpha\alpha$		$-d+j$	$d+j$	
ψ_+	$-d+j$			$d-j$
ψ_-	$d+j$			$-d-j$
$\beta\beta$		$d-j$	$-d-j$	

A blank entry denotes that the transition is forbidden.
$s = \delta B/2, d = \Delta/2, j = J/2$.

Problem 4.4.8.   *Show that if all the non-zero matrix elements between the basis functions are equal to μ, the two matrix elements between ψ_+ and $|\alpha\alpha>$ and $|\beta\beta>$ are equal to $(\cos\theta + \sin\theta)\mu$. Similarly show that the non-zero matrix elements involving ψ_- are $(\cos\theta - \sin\theta)\mu$. The transition intensities are therefore proportional to $1 + \sin 2\theta$ and $1 - \sin 2\theta$ respectively.*

Problem 4.4.9.   *In the case of two equivalent nuclei $\delta = 0$ show that $\theta = \pi/4$ and that the frequencies and intensities of the NMR spectrum reduce to those derived in the last subsection.*

Typical spectra are shown for different values of θ in Fig. 4.9. In each case a pattern symmetrical about the mean unperturbed chemical shift is found. The two central lines are separated by $\Delta - J$ and have intensities proportional to $1 + \sin 2\theta$. Each outer line is separated from the nearest central line by J and has an intensity proportional to $1 - \sin 2\theta$.

Problem 4.4.10.   *In the other limiting case $\delta B \gg J$ show that θ approaches zero and the normal first-order AX spectrum is obtained.*

4.5   Atoms in magnetic fields

When an atom is placed in an external magnetic field, the electronic wavefunction is perturbed by the field. As we have seen in Chapter 1 and Section 4.3, in the absence of a magnetic field the state of an atom can be

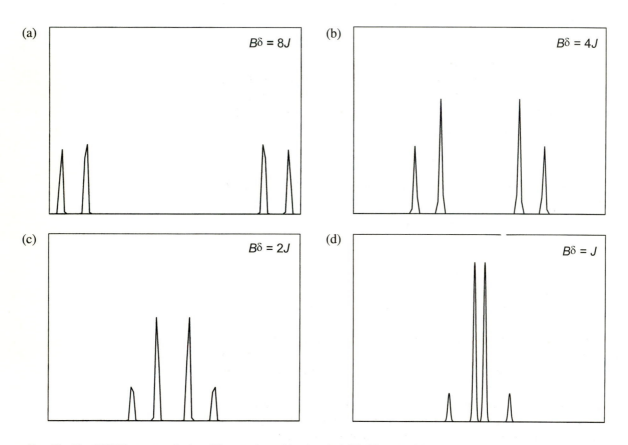

(a) $B\delta = 8J$

(b) $B\delta = 4J$

(c) $B\delta = 2J$

(d) $B\delta = J$

Fig. 4.9   The AB NMR spectrum for four different values of the chemical shift difference. $B\delta = J$, $2J$, $4J$, and $8J$.

characterized in the first instance by specifying the electron configuration together with the total orbital angular momentum L and the total spin angular momentum S. There is a weak interaction between the spin and orbital angular momenta, spin–orbit coupling, which results in small energy differences between states with the same configuration, L and S, but different values of the overall total electronic angular momentum J. Spin–orbit coupling is conveniently treated as a perturbation; however, in the presence of an external magnetic field the spin and the orbital angular momentum respond differently: the Hamiltonian for the interaction is

$$\hat{H}_z = \frac{B\mu_B}{\hbar} \left(\sum_i \hat{l}_z + 2 \sum_i \hat{s}_z \right) \tag{4.59}$$

$$= \frac{B\mu_B}{\hbar} \left(\hat{L}_z + 2\hat{S}_z \right) \tag{4.60}$$

The factor of 2 multiplying the second term is the g factor for electron spin. Because of this factor, the perturbation cannot be expressed solely in terms of $\hat{J}_z$ and $\hat{J}^2$, the most appropriate description depends on which of the two perturbations is the strongest. As in the case of the two-proton spectrum in NMR we identify two limiting cases.

Strong fields – the Paschen–Back effect

If the field effect is greater than the spin–orbit coupling, it is appropriate to choose a basis where the interaction Hamiltonian is diagonal, that is, the basis whose states are defined by the quantum numbers L, M_L, S, and M_S. The first-order energy shift is then simply given by

$$E_z^{(1)} = B\mu_B(M_L + 2M_S) \tag{4.61}$$

The small correction for the spin–orbit coupling can be calculated as the expectation:

$$E_{so}^{(1)} = A < LM_LSM_S|\hat{\mathbf{L}}.\hat{\mathbf{S}}|LM_LSM_S > \tag{4.62}$$

Problem 4.5.1. *Show that the energy level formula just derived, in conjunction with the selection rules $\Delta M_S = 0$ and $\Delta M_L = 0, \pm 1$ predicts that in high field any atomic transition should be split into a multiplet of three lines with a splitting of $B\mu_B$. This effect is known as the Paschen–Back effect.*

As before, the dot product could be expressed in terms of ladder operators, but in first order only the diagonal term $\hat{L}_z\hat{S}_z$ contributes to the expectation. The off-diagonal terms result in first-order corrections to the wavefunction, and therefore, only higher order corrections to the energy. The classical interpretation of this effect is that the two angular momenta $\mathbf{L}$ and $\mathbf{S}$ *precess* rapidly about the magnetic field, so that the vectors average out to L_z and S_z. On the slower timescale of the spin–orbit coupling only these average components are meaningful, so in the first order $\mathbf{L}.\mathbf{S}$ is given by L_zS_z. This

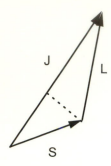

Fig. 4.10 The vector addition in spin–orbit coupling, illustrating the component of **S** in the direction of **J**

discussion is very similar to the analysis of the AX-type NMR spectrum. The final result for high fields is therefore

$$E^{(1)} = B\mu_B(M_L + 2M_S) + AM_LM_S \qquad (4.63)$$

Weak fields – the Lande g factor

In the absence of a field we have seen that the appropriate description of an atomic state is in terms of the quantum numbers L, S, and J. If a weak field is applied the additional quantum number M_J comes into play, as discussed in Section 4.2, but the quantum numbers M_L and M_S are no longer well defined. This is because the coupling of **L** and **S** to form **J** is much stronger than the interaction with the field. In the classical vector model the vectors **L** and **S** precess about their resultant, **J**, so that the only components that survive to interact with the external field are the components parallel to **J**.

The component of **S** parallel to **J** is found to be $(\mathbf{S.J})\mathbf{J}/J^2$, using a simple vector calculation illustrated in Fig. 4.10. A similar equation holds for the component of **L** in the direction of **J**. The magnetic field interaction can therefore be represented by the Hamiltonian

$$\hat{H}_z = \frac{B\mu_B}{\hbar}\left(\frac{\hat{\mathbf{L}}.\hat{\mathbf{J}} + 2\hat{\mathbf{S}}.\hat{\mathbf{J}}}{\hat{J}^2}\right)\hat{J}_z = \frac{B\mu_B}{\hbar}\left(1 + \frac{\hat{\mathbf{S}}.\hat{\mathbf{J}}}{\hat{J}^2}\right)\hat{J}_z \qquad (4.64)$$

Problem 4.5.2. *The strongest lines in the spectrum of the sodium atom have been assigned to transitions from the ground state $^2S_{1/2}$ to the first excited configuration, which has energy levels with term symbols $^2P_{1/2}$ and $^2P_{3/2}$. Use the Lande g-factor to predict how the two lines will be split in a weak magnetic field. This splitting is known as the anomalous Zeeman effect.*

We can use the same trick as we discussed in Section 4.3 to represent the dot product in terms of operators we know about

$$\hat{\mathbf{S}}.\hat{\mathbf{J}} = \frac{1}{2}(\hat{J}^2 - \hat{L}^2 + \hat{S}^2) \qquad (4.65)$$

Substituting these expressions into the Hamiltonian we see that the first-order energy shifts are

$$E_z^{(1)} = B\mu_B M_J\left(1 + \frac{J(J+1) - L(L+1) + S(S+1)}{2J(J+1)}\right) \qquad (4.66)$$

Problem 4.5.3. *Show that in a singlet to singlet transition in an atomic spectrum the normal Zeeman effect is observed, i.e. that the transitions split into three with splitting $B\mu_B$, as in the Paschen-Back effect.*

The factor in brackets is known as the *Lande g factor*.

In the magnetic field each energy level $|LSJ>$ is therefore split into its constituents according to the value of the magnetic quantum number M_J. In an atomic spectrum these splittings in the energy levels give rise to splittings in the absorption and emission lines, which are subject to the selection rule $\Delta M_J = 0, \pm1$.

References

Cohen-Tannoudji, C., Diu, B. and Laloë, F. (1992). *Mécanique quantique II*, Chapter 12. Hermann, Paris.

Green, N. J. B. (1997). *Quantum mechanics 1*. Oxford Chemistry Primer, Oxford.

Herzberg, G. (1950). *Spectra of diatomic molecules*. Van Nostrand Reinhold, New York.

Landau, L. D. and Lifschitz, E. M. (1977). *Quantum mechanics*, 3rd edn. Pergamon, Oxford.

Pauling, L. and Wilson, E. B. (1985). *Introduction to quantum mechanics*. Dover, New York.

5 Time

5.1 The time-dependent Schrödinger equation

Time does not appear in a consistent or natural way in the Schrödinger theory.

Up to this point we have made no attempt to use quantum mechanics to describe time-dependent phenomena, even though much of chemistry deals with dynamics on a molecular scale; the kinetics of chemical reaction, the motion and collision of molecules are obvious examples. However, the position of time in the Schrödinger theory is not wholly satisfactory. Time is not treated as an observable in the same way as position or momentum, but as a parameter, which can be used to follow the evolution of an event. Time is not therefore represented by an operator.

However, this view is incomplete: in the theory of relativity time becomes a fourth spatial dimension. Changing frame of reference is a linear transformation of coordinate system in four-dimensional space–time. Quantum mechanics should show this symmetry by treating time in the same way as the three spatial dimensions so that transformations between different frames of reference can be made in a consistent way. This is not possible in the Schrödinger theory, a defect rectified in the relativistic theory of Dirac, which is beyond the scope of this book (see Dirac 1958; Bjorken and Drell 1964; Davydov 1976).

In spite of this defect, classical (non-relativistic) mechanics provides an accurate description of macroscopic systems unless speeds approach the magnitude of the speed of light. In the same way, the Schrödinger time-dependent theory provides an accurate description of transient phenomena on the microscopic scale as long as speeds remain small.

The best way of deriving the time-dependent Schrödinger equation is to start with the fully relativistic Dirac theory and take the appropriate limit. This has the virtue of introducing both time and spin in a proper way. However, the same result can be obtained in a much more physically intuitive way, by returning to the description of a free particle travelling in one dimension with a momentum p. The wavefunction of such a particle was given in *Quantum mechanics 1*, Section 1.3 as

$$\psi(x) = \exp(2\pi i x/\lambda) \tag{5.1}$$

where λ is the de Broglie wavelength. Since this wavefunction represents a particle travelling through space with momentum $p = h/\lambda$, it is natural to attempt to include time in such a way that the wavefunction becomes a travelling wave in the positive x direction:

$$\Psi(x, t) = \exp[2\pi i(x/\lambda - vt)] \tag{5.2}$$

If we take a snapshot of the wavefunction at a particular time we can see that real and imaginary parts vary sinusoidally with a wavelength λ. Alternatively, if we sit at a particular point and watch the wave go past we see that it oscillates sinusoidally in time with a frequency v. The energy of the particle is given by the usual formula $E = p^2/2m$; we postulate that this should be consistent with Planck's law $E = hv$; equating the two and using the de Broglie relation we find that eqn 5.2 can be rewritten as

$$\Psi(x, t) = \exp[\mathrm{i}(px - Et)/\hbar] \qquad (5.3)$$

It we differentiate Ψ with respect to time and multiply by $\mathrm{i}\hbar$

$$\mathrm{i}\hbar\frac{\partial\Psi}{\partial t} = E\Psi \qquad (5.4)$$

The time derivative operator has the same effect on the wavefunction of a free particle as the Hamiltonian. In the Schrödinger theory this relationship is assumed to be general, leading to the time-dependent Schrödinger equation:

$$\mathrm{i}\hbar\frac{\partial\Psi}{\partial t} = \hat{H}\Psi \qquad (5.5)$$

The form of the time derivative operator and the appearance of the energy in the imaginary exponent of the wavefunction are strongly suggestive that the energy has the properties of the momentum conjugate to t, and indeed in the relativistic Dirac theory the energy operator does have this interpretation (Dirac 1958; Davydov 1976).

The wavelength and frequency of a quantum mechanical matter wave are related by

$$v = h/2m\lambda^2 \qquad (5.6)$$

This equation differs from the corresponding equation for light by a factor of 2 because the kinetic energy of a free particle is equal to $mv^2/2$, whereas the energy of a photon is mc^2.

5.2 Stationary states

The time-dependent Schrödinger equation is the quantum mechanical analogue of Hamilton's equation of motion in classical mechanics. Because of this relationship wavefunctions with an exactly defined energy (eigenfunctions of the Hamiltonian) have a particularly simple dependence on time, corresponding to the classical observation that the energy of an isolated system is a constant of the motion. This may be seen by applying the separation of variables method (*Quantum mechanics 1*, Chapter 2) to eqn 5.5. (In this chapter we use ψ for a time-independent wavefunction and Ψ for a time-dependent wavefunction.)

Searching for a solution of the form $T(t)\psi(\mathbf{r})$ we arrive at the separated equations

$$\hat{H}\psi = E\psi$$
$$\frac{\mathrm{d}T}{\mathrm{d}t} = \frac{-\mathrm{i}E}{\hbar}T \qquad (5.7)$$

where E is the separation constant. Such a solution exists if ψ is an eigenfunction of the Hamiltonian, that is, a solution of the time-independent Schrödinger equation, and E is the energy, in which case the solution to the time-dependent Schrödinger equation is given by

$$\Psi(\mathbf{r}, t) = \psi(\mathbf{r})\exp(-\mathrm{i}Et/\hbar) \qquad (5.8)$$

As long as the Hamiltonian itself does not depend on time, the spatial part of the wavefunction remains constant and, in consequence, so does the energy; just as in classical mechanics, the energy is a constant of the motion. For this reason eigenfunctions of the Hamiltonian are referred to as *stationary states*.

When the Hamiltonian is time-independent, these solutions can be used in conjunction with the method of expansion in eigenfunctions (*Quantum mechanics 1*, Section 1.5) to describe the time evolution of any wavefunction. Let the eigenfunctions of $\hat{H}$ be represented as kets, so that, for example, $\hat{H}|j> = E_j|j>$, and suppose that at time zero the system of interest has wavefunction ψ, which can be expanded,

$$\Psi(\mathbf{r}, 0) = \psi(\mathbf{r}) = \sum_j c_j|j> \tag{5.9}$$

To describe the time-dependence of the system it is simply necessary to multiply each term in the summation by $\exp(-iE_j t/\hbar)$, using the appropriate energy. Each term will then obey eqn 5.5 separately, hence so will any superposition of them:

$$\Psi(\mathbf{r}, t) = \sum_j c_j \exp(-iE_j t/\hbar)|j> \tag{5.10}$$

We also note that the probability of finding the system in a particular stationary state, $|k>$, is time-independent:

$$P_k = c_k^* c_k \times \exp(iE_k t/\hbar)\exp(-iE_k t/\hbar) = c_k^* c_k \tag{5.11}$$

because the two complex conjugate exponentials cancel out. Thus, if the Hamiltonian is time-independent the system is stationary in the sense that the probability distribution of the energy is independent of time.

Problem 5.2.1. *Prove by direct substitution that eqn 5.10 is a solution of the time-dependent Schrödinger equation.*

Problem 5.2.2. *Show that the expectation energy is time-independent.*

Problem 5.2.3. *Show that the normalization of Ψ is time-independent. (Hint: $\hat{H}$ is Hermitian, so that its eigenfunctions are orthogonal.)*

5.3   Equations of motion

However, other observable physical properties are not necessarily time-independent. Let us investigate the expectation of some observable A, represented by the time-independent operator $\hat{A}$. Differentiating under the integral sign, and using the product rule, we find:

$$\frac{d<A>}{dt} = \int \frac{\partial \Psi^*}{\partial t}\hat{A}\Psi \, d\tau + \int \Psi^* \frac{\partial \hat{A}}{\partial t}\Psi \, d\tau + \int \Psi^*\hat{A}\frac{\partial \Psi}{\partial t} \, d\tau \tag{5.12}$$

The second term is zero if the operator $\hat{A}$ is independent of time. In the first and third terms we use the time-dependent Schrödinger equation for the time derivative to obtain

$$\frac{d<A>}{dt} = \frac{i}{\hbar}\int (\hat{H}^*\Psi^*)\hat{A}\Psi \, d\tau - \frac{i}{\hbar}\int \Psi^*\hat{A}\hat{H}\Psi \, d\tau$$

Since $\hat{H}$ is Hermitian we can simplify the first term

$$\int (\hat{H}^*\Psi^*)\hat{A}\Psi \, d\tau = \int \Psi^*\hat{H}\hat{A}\Psi \, d\tau$$

leaving the result

$$\frac{\mathrm{d} <A>}{\mathrm{d}t} = \frac{\mathrm{i}}{\hbar} \int \Psi^*(\hat{H}\hat{A} - \hat{A}\hat{H})\Psi \, \mathrm{d}\tau \qquad (5.13)$$

If the operator $\hat{A}$ commutes with the Hamiltonian the quantity A is said to be *conserved* because its expectation (and its probability distribution) are time-independent. Classically, such a quantity is said to be a constant of the motion.

The results of the two problems are the quantum mechanical generalization of the classical equations of motion of the particle. The first says that the rate of change of position is the velocity, and the second is essentially the generalization of Newton's second law, force (minus gradient of potential energy) = rate of change of momentum, and is often known as *Ehrenfest's theorem*.

Problem 5.3.1. *Show that the time-derivative of the expectation x coordinate of a particle is the x component of the expectation velocity.*

Problem 5.3.2. *Show that the rate of change of the expectation momentum in the x direction is given by*

$$\frac{\mathrm{d} <p_x>}{\mathrm{d}t} = -\left\langle \frac{\partial V}{\partial x} \right\rangle$$

where V(x) is the potential energy.

A two-state system

As a simple example of a time-dependent phenomenon let us consider a system with two states. The system we shall consider is a pair of free radicals in a strong magnetic field. The two radicals are different chemical species, and their energies are modified by the field according to the Zeeman effect (Section 4.2), so that the energy of the pair becomes (see eqn 4.12)

$$E(m_1, m_2) = (g_1 m_1 + g_2 m_2)\mu_B B \qquad (5.14)$$

In particular, the states $|\alpha\beta>$ and $|\beta\alpha>$ have the energies $\pm\hbar\omega/2$, where $\hbar\omega = (g_1 - g_2)\mu_B B$. Suppose that the radical pair is formed by flash photolysis in a singlet state, which is a superposition of the two states $|\alpha\beta>$ and $|\beta\alpha>$:

$$|S> = \frac{1}{\sqrt{2}}(|\alpha\beta> - |\beta\alpha>) \qquad (5.15)$$

and recall that the triplet state $|T_0>$ is a similar superposition, with a plus sign in place of the minus sign.

If the initial wavefunction is $|S>$, we can immediately write down the wavefunction at time t from eqn 5.10:

$$\Psi(t) = \frac{1}{\sqrt{2}}(e^{-\mathrm{i}\omega t/2}|\alpha\beta> - e^{\mathrm{i}\omega t/2}|\beta\alpha>) \qquad (5.16)$$

However, there are two non-commuting observables in this system. The energy, which is conserved, as discussed above, and the total spin quantum number of the pair, which determines its singlet or triplet nature. If we express Ψ in terms of $|S>$ and $|T_0>$ we can see this more clearly

$$\Psi(t) = \frac{1}{2}(e^{-\mathrm{i}\omega t/2}(|T_0> + |S>) - e^{\mathrm{i}\omega t/2}(|T_0> - |S>))$$

$$= \cos\frac{\omega t}{2}|S> - \mathrm{i}\sin\frac{\omega t}{2}|T_0> \qquad (5.17)$$

The singlet probability is therefore $\frac{1}{2}(1 + \cos\omega t)$, and the triplet probability is $\frac{1}{2}(1 - \cos\omega t)$. We can see that the singlet probability oscillates coherently

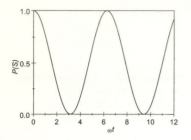

Fig. 5.1 The coherent oscillation of the singlet probability of a radical pair in a magnetic field calculated from eqn 5.17.

A wavepacket is a wavefunction prepared in such a way that it is localized in space. It is a superposition of several stationary states with well defined phase relationships.

between 1 and 0 with a period $2\pi/\omega$, as shown in Fig. 5.1. This is an example of a phenomenon known as *Rabi oscillation*.

The singlet probability is an observable because when two radicals encounter, their fate depends on this spin state. For example, in some cases only the singlet state leads to reaction – then the singlet probability corresponds to the probability of reaction on encounter. In other cases reaction leads to a singlet state, which fluoresces, or to a triplet state, which does not. Either way, the rate of formation of the singlet state oscillates coherently. Such oscillations have been observed experimentally in several systems, and are known as *quantum beats*.

Wavepackets

Many important modern experiments prepare systems in such a way that the initial wavefunction is not an eigenstate of the Hamiltonian. If this initial wavefunction is localized in space in some way, the terms that appear in eqn 5.10 must have a definite phase relationship so that they interfere constructively in the region where the particle is localized and destructively elsewhere. Such a state is often called a wavepacket to emphasize its localization. A wavepacket does not necessarily remain localized, it will generally tend to spread out as time evolves, but its time evolution is entirely determined by the different rates at which the phases of each term in the expansion increase, following eqn 5.10.

Several types of experiment in modern chemistry produce wavepackets. The best known is the vibrational wavepacket, in which a molecule undergoes an electronic transition into a superposition of excited vibrational states of the excited potential energy surface, determined by the Franck–Condon factors. In a diatomic molecule the wavepacket produced in this way oscillates back and forth in the potential well in a pseudo-classical way, gradually spreading out and producing interference effects. If the curve crosses a dissociative state then every time the wavepacket passes through the crossing region a part of it follows the dissociative curve and is lost. The rest remains bound. Another example is a pulse of light in a mode-locked laser, which is a superposition of the longitudinal modes (standing waves) of the laser cavity, and bounces back and forth between the mirrors at the ends of the cavity.

Similar experiments can be performed with electronic wavepackets, that is, superpositions of electronic wavefunctions, by exciting an atom or molecule very close to its ionization limit, where the energy levels obey the Rydberg formula and are very closely spaced so that several states can be excited simultaneously and coherently. The wavepacket is produced in the core region and then orbits in a semi-classical manner, following the appropriate elliptical Sommerfeld orbit. The wavepacket spreads out as it orbits, and eventually the front end catches the back end and interference results. Every time the wavepacket passes through the core region a small part of it autoionizes and can be detected. In this way it is possible to watch electrons orbit in real time.

These experiments may seem simply to be curiosities, but they illustrate how many photochemical reactions occur. Absorption of light produces a wavepacket in the Franck–Condon footprint on the excited state potential energy surface; the subsequent motion of this vibrational wavepacket determines the chemical rearrangement of the nuclei. In principle this is a very powerful method for analysing photochemical reactions, but at present it is limited to small molecules because of the large number of dimensions (degrees of freedom) required for larger molecules. Larger molecules are generally dealt with, either by reducing the number of dimensions in the calculation, or by propagating classical trajectories rather than wavepackets. It is likely, however, that this area will see substantial progress in the next few years.

5.4 Sudden changes

So far we have dealt with phenomena where the Hamiltonian is constant in time. There are many physical phenomena in which the Hamiltonian depends on time. We shall start by considering the limiting case where $\hat{H}$ changes very rapidly, and permanently, to a new form; for example when a nucleus undergoes radioactive decay in an atom its charge is altered permanently, altering the motion of the electrons in the atom. Another example of a sudden change occurs when a molecule undergoes an electronic transition; the rearrangement of the electrons leads to a change in the potential energy function for the nuclear motion, and so affects the molecular vibrations.

In a sudden change the Hamiltonian changes very rapidly and the wavefunction does not have time to respond.

In order to analyse a change of this type, consider a system in which the Hamiltonian is $\hat{H}_0$, and the wavefunction is given by a stationary state ψ_k. At time zero the Hamiltonian rapidly and irreversibly changes to $\hat{H}$. If the change takes place sufficiently rapidly, the wavefunction does not have time to respond, and the system finds itself immediately after time zero with an unaltered wavefunction ψ_k, which is no longer a stationary state of the new Hamiltonian. This is the essential physical content of the *sudden approximation*.

However, the wavefunction ψ_k can be expressed as an expansion in terms of the eigenfunctions of the new Hamiltonian, $\hat{H}$,

$$\psi_k = \sum a_{kj}\phi_j \tag{5.18}$$

and following the method of *Quantum mechanics 1*, Section 1.5, the coefficients in the summation are given by

$$a_{kj} = \int \phi_j^* \psi_k \, d\tau \tag{5.19}$$

Since the Hamiltonian does not alter after time zero, the amplitude for the system to occupy state ϕ_j at some subsequent time t is given by $a_{kj}\exp(-iE_j t/\hbar)$, and so the probability of finding the system in this state is stationary and equal to $|a_{kj}|^2$.

Radioactive decay

As an example of the sudden approximation let us consider what happens to a tritium atom in its ground state when the nucleus undergoes a beta decay, and becomes a nucleus of ^{3}He. Explicitly, let us calculate the probability that the resulting helium ion will be found in its ground state, in which the electron occupies the $1s$ orbital, on the assumption that the radioactive decay takes place instantly on an electronic timescale.

The initial wavefunction is the $1s$ orbital of a hydrogen atom and is given by (in atomic coordinates)

$$\psi = \frac{1}{\sqrt{\pi}} e^{-r} \tag{5.20}$$

Problem 5.4.1. *Explain why, if the tritium atom is in its ground state, the probability of forming the helium ion with the electron in a p orbital is exactly zero. (Similarly for any type of orbital other than an s orbital.)*

We wish to calculate the probability that the final state is the $1s$ orbital of a He^+ ion, which is given by

$$\phi = \frac{2\sqrt{2}}{\sqrt{\pi}} e^{-2r} \tag{5.21}$$

Substituting these wavefunctions into the integral for the expansion coefficient a_{1s1s}, we find

$$a_{1s1s} = \frac{2\sqrt{2}}{\pi} 4\pi \int_0^\infty r^2\, e^{-3r}\, dr = \frac{16\sqrt{2}}{27} \tag{5.22}$$

Hence the probability of forming the ion in its ground electronic state is

$$|a_{1s1s}|^2 = \frac{512}{729} = 0.8381\ldots \tag{5.23}$$

The Franck–Condon principle

When an electronic transition takes place in a molecule the electron distribution is rearranged. It is this electron distribution which holds the molecule together and defines the potential energy function for the molecular vibrations. Since electrons move on the order of 1000 times faster than nuclei, the electronic transition is effectively instantaneous on the timescale of nuclear motion, and therefore the sudden approximation is appropriate. Before the transition the molecule has vibrational wavefunction ψ_k and the transition is sufficiently fast that this wavefunction is not altered. After the transition the potential energy function (and therefore the vibrational Hamiltonian) is changed, thus ψ_k is no longer a stationary state.

According to the sudden approximation, however, the amplitude for finding the molecule in vibrational state ϕ_j is a_{kj}, defined by eqn 5.19, and the probability of transition to this vibrational state is therefore $|a_{kj}|^2$. In this context the coefficient a_{kj} is the overlap integral between the initial vibrational wavefunction, an eigenfunction of $\hat{H}_0$, and the final vibrational wavefunction, an eigenfunction of $\hat{H}$. The intensity of transition is weighted by the Franck–Condon factor, which is the square modulus of the overlap integral, as previously discussed in *Quantum mechanics 1*, Section 2.5.

The discussion of this section shows clearly the relationship between the conventional derivation of the Franck–Condon principle and the idea of a sudden change in the potential energy. The Franck–Condon principle is often explained in classical terms, by saying that the sudden change in the potential energy function means that the nuclei suddenly find themselves in a nonequilibrium configuration, therefore the bond is either compressed or stretched and responds by vibrating, like a spring. This explanation is at best only a classical analogy – we can see from the analysis here that the sudden change in the potential energy does not lead to any change in the wavefunction, and to find the distribution of new vibrational states we simply expand the wavefunction in a series of eigenfunctions of the new Hamiltonian. The expansion coefficients are the overlap integrals, which can be either positive or negative and do not correspond to the probability of finding the bond at a particular extension.

5.5 Time-dependent perturbations

In the remainder of this chapter we consider systems in which the Hamiltonian is time-dependent. There are many examples of such systems. One of the most important in chemistry is the interaction of a molecule with light, another is the collision of two molecules, or an electron and a molecule. In each of these cases the external perturbation acts for some period of time and the Hamiltonian then returns to a constant form. The action of the perturbation is to induce transitions between the stationary states of the original system.

Let us assume that the time-dependent Hamiltonian can be split into two parts, one of which is constant, $\hat{H}_0$, and the other contains the time-dependent perturbation, $\hat{V}(t)$.

$$i\hbar \frac{\partial \Psi}{\partial t} = (\hat{H}_0 + \lambda \hat{V}(t))\Psi \tag{5.24}$$

As in Chapter 3, since a perturbation method is used, the parameter λ is introduced to keep track of the level of approximation. $\hat{V}$ is assumed to be small. The stationary states of $\hat{H}_0$, with energy E_j, will be denoted $\Psi_j = \exp(-iE_j t/\hbar)|j>$. In particular, at time zero the system will be assumed to occupy a stationary state Ψ_i. The normal method for attempting a solution of the full time-dependent Schrödinger equation is to express the unknown solution as an expansion in the known stationary states of the unperturbed equation:

$$\Psi_i = \sum_j c_{ji}(t) \exp(-iE_j t/\hbar)|j> \tag{5.25}$$

Substituting this summation into the time-dependent Schrödinger equation, we find

$$i\hbar \sum_j e^{-iE_j t/\hbar}|j> \frac{dc_{ji}}{dt} = \sum_j e^{-iE_j/\hbar} c_{ji}\lambda \hat{V}|j> \tag{5.26}$$

Premultiplying both sides by Ψ_k^* and integrating over all space, only one term of the summation on the left hand side survives (because the Ψ_j are orthogonal), giving

$$i\hbar \frac{dc_{ki}}{dt} = \sum_j e^{i\omega_{kj}t/\hbar} c_{ji}\lambda <k|\hat{V}|j> \tag{5.27}$$

where $\hbar\omega_{kj} = E_k - E_j$. The set of coupled differential equations derived in this way is an exact transformation of the Schrödinger equation. In many applications the mathematical analysis would stop here, since there are standard numerical routines for the computational solution of systems of this form. However, a great deal can be learned from an approximate analysis using perturbation theory. We therefore search for a solution of the form

$$c_{ki} = c_{ki}^{(0)} + \lambda c_{ki}^{(1)} + \dots \tag{5.28}$$

and obtain a set of equations for each order of approximation in terms of the previous order:

$$i\hbar\frac{dc_{ki}^{(0)}}{dt} = 0 \tag{5.29}$$

$$i\hbar\frac{dc_{ki}^{(1)}}{dt} = \sum_j e^{i\omega_{kj}t/\hbar} c_{ji}^{(0)} < k|\hat{V}|j > \tag{5.30}$$

$$\dots$$

The unperturbed wavefunction is Ψ_i, so in the zero-order approximation we can write $c_{ii}^{(0)} = 1$ and $c_{ki}^{(0)} = 0$. The first-order approximation is now given by

$$i\hbar\frac{dc_{ki}^{(1)}}{dt} = e^{i\omega_{ki}t/\hbar} < k|\hat{V}|i > \tag{5.31}$$

which can be integrated to give

$$i\hbar c_{ki}^{(1)} = \int_0^t e^{i\omega_{ki}t/\hbar} < k|\hat{V}|i > \, dt \tag{5.32}$$

This completes the development of the first-order solution.

Very slow changes

If the Hamiltonian changes very slowly the system remains in a stationary state throughout. No transitions are excited. Such a transition is called *adiabatic*.

We notice that if the Hamiltonian changes from one form to another, then the integral in eqn 5.32 cannot be easily evaluated because it varies sinusoidally at long times. This problem can be remedied by an integration by parts:

$$c_{ki}^{(1)} = -\left[\frac{e^{i\omega_{ki}t/\hbar} < k|\hat{V}|i >}{\hbar\omega_{ki}}\right]_0^t + \int_0^t \frac{e^{i\omega_{ki}t/\hbar}}{\hbar\omega_{ki}} < k|\frac{\partial\hat{V}}{\partial t}|i > \, dt \tag{5.33}$$

The first term vanishes at $t = 0$, where the perturbation is zero, and at long times the upper limit becomes the first-order correction to the final wavefunction (see Chapter 3). The first term does not, therefore, induce transitions, but describes how a stationary state of the initial Hamiltonian evolves into the corresponding stationary state of the perturbed Hamiltonian.

The second term contains all the information about transitions induced by the perturbation. The transition probability is given by the square modulus,

$$P_{i \to k} = \left| \int_0^t \frac{e^{i\omega_{ki}t/\hbar}}{\hbar\omega_{ki}} < k|\frac{\partial\hat{V}}{\partial t}|i > \, dt \right|^2 \tag{5.34}$$

If the Hamiltonian changes little on timescales of the order of $1/\omega_{ki}$, the sinusoidally varying function $\exp(i\omega_{ki}t)$ will pass through many oscillations before the Hamiltonian changes appreciably. In this case the integral will be close to zero and no transitions will take place. Such a change is called *adiabatic*. In an adiabatic change the Hamiltonian changes so slowly that the wavefunction effectively remains in a stationary state, and no transitions are induced to other states. This is the opposite of the sudden approximation. Many theories of chemical kinetics rely on the adiabatic approximation, since they treat chemical reaction in terms of nuclear motion along a single potential energy curve or surface.

Absorption of light

The final important type of time-dependent perturbation is found when the perturbation itself fluctuates sinusoidally. This type of perturbation is characteristic of the interaction of a molecule with light, leading to the absorption or emission of electromagnetic radiation. Such a perturbation is typically of the form

$$\hat{V}(t) = \hat{W} \sin \omega t \qquad (5.35)$$

where $\hat{W}$ is independent of time. In absorption spectroscopy $\hat{W}$ is proportional to the component of the dipole moment operator in the direction of the electric vector of the light.

Applying eqn 5.32 we find the first-order coefficient

$$c_{ki} = \frac{<k|\hat{W}|i>}{2i\hbar} \left[\frac{1 - e^{i(\omega_{ki}+\omega)t}}{\omega_{ki} + \omega} - \frac{1 - e^{i(\omega_{ki}-\omega)t}}{\omega_{ki} - \omega} \right] \qquad (5.36)$$

The absorption of light leading to excitation from state $|i>$ to state $|k>$ is not normally important unless the frequency is close to resonance with frequency ω_{ki}. Under these conditions $|\omega_{ki} - \omega| << |\omega_{ki}|$ and the first term of eqn 5.36 is negligible in comparison to the second term. We therefore have

$$c_{ki} = -\frac{<k|\hat{W}|i>}{2i\hbar} \frac{1 - e^{i(\omega_{ki}-\omega)t}}{\omega_{ki} - \omega}$$

$$= \frac{<k|\hat{W}|i>}{2\hbar} e^{i(\omega_{ki}-\omega)t/2} \frac{\sin[(\omega_{ki} - \omega)t/2]}{(\omega_{ki} - \omega)/2} \qquad (5.37)$$

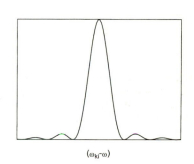

$(\omega_{ki}\text{-}\omega)$

Fig. 5.2 The frequency dependence of the absorption probability according to the golden rule.

leading to the transition probability known as the *golden rule* (Fig. 5.2):

$$P_{i\to k} = \frac{|<k|\hat{W}|i>|^2}{\hbar^2} \frac{\sin^2[(\omega_{ki} - \omega)t/2]}{(\omega_{ki} - \omega)^2} \qquad (5.38)$$

The similar case where ω is close to $-\omega_{ki}$ leads to the phenomenon of stimulated emission.

The transition probability is the square modulus of this coefficient and is seen to be proportional to $|<k|\hat{W}|i>|^2$, the square modulus of the dipole moment matrix element. This is the origin of selection rules in spectroscopy, whose consequences have already been discussed, for example in *Quantum mechanics 1*, Section 4.8.

5.6   The quantum Zeno paradox

We finish the discussion of time-dependent phenomena with a curiosity, which illustrates very clearly the differences between the classical and quantum mechanical conceptions of probability. Let us return to a two-state system like the radical pair discussed in Section 5.3, whose stationary states are separated by an energy difference $\hbar\omega$. In terms of the eigenstates of some observable (e.g. M_S) which does not commute with the Hamiltonian these states are $(|0> \pm |1>)/\sqrt{2}$. This is the simplest possible example of a time-dependent system, and it acts as a model for many real physical phenomena, for example the radical pair of Section 5.3, the inversion of the ammonia molecule, resonance in the π bonding of the benzene molecule, the delocalization of a hole in a 1s orbital of the N_2^+ ion (see Feynman *et al.* 1965) for discussions of many such systems).

We have already seen in Section 5.3 that if the system is initially described by the wavefunction $|0>$, then the subsequent evolution of the wavefunction is given by

$$\Psi(t) = \cos(\omega t/2)|0> - \text{i}\sin(\omega t/2)|1> \qquad (5.39)$$

At time $T = \pi/\omega$ the wavefunction will be given by $-\text{i}|1>$, that is a measurement at time T is certain to find the particle in state $|1>$. Similarly at $2T$ the wavefunction has evolved coherently to $-|0>$, and a measurement is sure to find the system back in its original state. In the absence of any external perturbation this coherent oscillation will continue indefinitely.

However, let us now consider the effect of making a measurement at $T/2$. At this moment the wavefunction is

$$\Psi(T/2) = \frac{1}{\sqrt{2}}(|0> - \text{i}|1>) \qquad (5.40)$$

Let us further assume that the measurement is effectively instantaneous on the timescale of T, and that if the system is found to be in state $|0>$, by the measurement, then the wavefunction 'collapses' on to $|0>$, and similarly for state $|1>$. (Such idealized measurements do exist, for example the Stern–Gerlach experiment.) Once the measurement has been made, the wavefunction starts to evolve coherently again.

At $T/2$ the measurement is equally likely to find the system in each of the two states. But there is an important difference between the system before the measurement, which is a quantum mechanical superposition of the two states $|0>$ and $|1>$, and after the measurement, which is a classical probability, that is the system either has wavefunction $|0>$, with probability $\frac{1}{2}$ or it has wavefunction $|1>$ with probability $\frac{1}{2}$. If the state is found to be $|0>$, then at time T (that is, after a second period $T/2$) the wavefunction will again be given by eqn 5.40, and the probability of finding state $|1>$ is only $\frac{1}{2}$.

Similarly, if the state at $T/2$ was found to be $|1>$, the wavefunction at T would be

$$\Psi(T) = \frac{1}{\sqrt{2}}(-\mathrm{i}|0> + |1>) \tag{5.41}$$

and again, the probability of finding the system in $|1>$ is $\frac{1}{2}$. Combining these two possibilities we conclude that the effect of making an observation at $T/2$ is to reduce the probability of finding the system in $|1>$ at time T from 1 to $\frac{1}{2}$.

This analysis can be generalized. Let T be divided into n equal intervals, and let an observation be made at each division. If the previous measurement was $|0>$ the wavefunction at the next measurement is

$$\Psi(t) = \cos(\pi/2n)|0> -\mathrm{i}\sin(\omega\pi/2n)|1> \tag{5.42}$$

and the same equation applies with $|0>$ and $|1>$ reversed if the previous measurement was $|1>$.

Now let the probability that the kth measurement gives $|1>$ be denoted P_k. Then it is easy to see that

$$P_{k+1} = (1 - P_k)\sin^2(\pi/2n) + P_k\cos^2(\pi/2n) \tag{5.43}$$

which is a recurrence relationship for the P_k, whose solution is given by

$$P_k = \frac{1}{2}\left(1 - \cos^k\frac{\pi}{k}\right) \tag{5.44}$$

Problem 5.6.1. *Solve the recurrence relation for P_k to obtain eqn 5.44.*

But P_n is the probability of finding $|1>$ after n intervals, that is, at time T. We can see that for $n = 1$ the general formula gives a probability of 1, and for $n = 2$ we get $\frac{1}{2}$ as argued above. For $n = 3$ the probability is $\frac{7}{16}$, and as the number of divisions increases the probability of finding the system in state $|1>$ decreases towards zero in the limit $n \to \infty$.

Problem 5.6.2. *Prove that P_n vanishes in the limit $n \to \infty$.*

Thus we see that the more often we measure to see if we have reached state $|1>$, the less likely we are ever to get there. The disturbance induced by the act of making the measurement collapses the coherent quantum mechanical superposition of states into a classical probabilistic mixture of pure states, that is, it destroys the coherence (phase relationship) between the states; the effect of this change is to inhibit the transition from one state to another. This analysis gives a new twist to the observation 'a watched pot never boils'.

References

Bjorken, J. D. and Drell, S. D. (1964). *Relativistic quantum mechanics*. McGraw-Hill, New York.

Davydov, A. S. (1976). *Quantum mechanics*, 2nd edn. Pergamon Press, Oxford.

Dirac, P.\ A.\ M. (1958). *The principles of quantum mechanics*, 4th edn. Clarendon Press, Oxford.

Feynman, R. P., Leighton, R. B., and Sands, M. (1965). *The Feynman lectures on physics*, Vol. III. Addison-Wesley, Reading.

Green, N. J. B. (1997). *Quantum mechanics 1*. Oxford Chemistry Primer, Oxford.

Index